D0471077

LEARNING MATHEMATICS

Second edition

Available in the Cassell Education series

P. Ainley: *Young People Leaving Home*
P. Ainley and M. Corney: *Training for the Future: The Rise and Fall of the Manpower Services Commission*
G. Antonouris and J. Wilson: *Equal Opportunities in Schools*
L. Bash and D. Coulby: *The Education Reform Act: Competition and Control*
N. Bennett and A. Cass: *From Special to Ordinary Schools*
D. E. Bland: *Managing Higher Education*
M. Booth, J. Furlong and M. Wilkin: *Partnership in Initial Teacher Training*
M. Bottery: *The Morality of the School*
L. Burton (ed.): *Gender and Mathematics*
C. Christofi: *Assessment and Profiling in Science*
G. Claxton: *Being a Teacher: A Positive Approach to Change and Stress*
G. Claxton: *Teaching to Learn: A Direction for Education*
D. Coulby and L. Bash: *Contradiction and Conflict: The 1988 Education Act in Action*
D. Coulby and S. Ward (eds): *The Primary Core National Curriculum*
C. Cullingford (ed.): *The Primary Teacher*
L. B. Curzon: *Teaching in Further Education (4th edition)*
P. Daunt: *Meeting Disability: A European Response*
J. Freeman: *Gifted Children Growing Up*
B. Goacher *et al.*: *Policy and Provision for Special Educational Needs*
H. L. Gray (ed.): *Management Consultancy in Schools*
L. Hall: *Poetry for Life*
J. Nias, G. Southworth and R. Yeomans: *Staff Relationships in the Primary School*
A. Orton: *Learning Mathematics: Issues, Theory and Classroom Practice (2nd edition)*
A. Pollard: *The Social World of the Primary School*
R. Ritchie (ed.): *Profiling in Primary Schools*
J. Sayer and V. Williams (eds): *Schools and External Relations: Managing the New Partnerships*
B. Spiecker and R. Straughan (eds): *Freedom and Indoctrination in Education: International Perspectives*
R. Straughan: *Beliefs, Behaviour and Education*
M. Styles, E. Bearne and V. Watson (eds): *After Alice: Exploring Children's Literature*
S. Tann: *Developing Language in the Primary Classroom*
H. Thomas: *Education Costs and Performance*
H. Thomas, with G. Kirkpatrick and E. Nicholson: *Financial Delegation and the Local Management of Schools*
D. Thyer and J. Maggs: *Teaching Mathematics to Young Children (3rd edition)*
M. Watts: *Science in the National Curriculum*
M. Watts: *The Science of Problem-solving*
J. Wilson: *A New Introduction to Moral Education*
S. Wolfendale (ed.): *Parental Involvement*

Learning Mathematics

Issues, Theory and Classroom Practice

Second edition

Anthony Orton

CASSELL

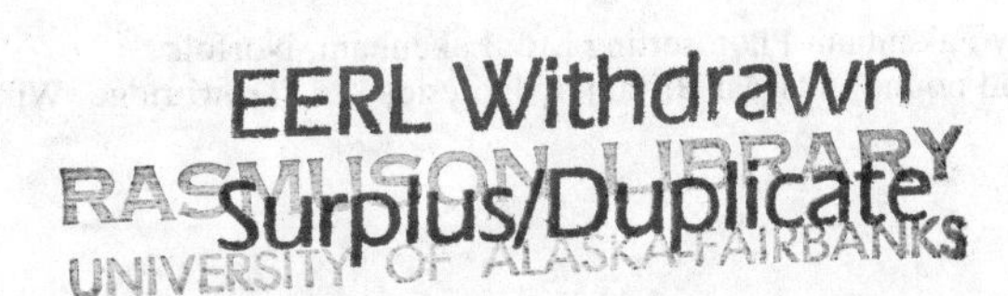

Cassell
Villiers House
41/47 Strand
London WC2N 5JE

387 Park Avenue South
New York, NY 10016–8810
USA

First published 1987. Second edition 1992.

British Library Cataloguing-in-Publication Data
A catalogue record for this book is available from the British Library

Library of Congress Cataloging-in-Publication Data

Orton, Anthony.
Learning mathematics: issues, theory and classroom practice / Anthony Orton—2nd ed.
p. cm.—(Cassell education series)
Includes bibliographical references and index.
ISBN 0–304–32553–8—ISBN 0–304–32555–4 (pbk.)
1. Mathematics—Study and teaching. I. Title. II. Series: Cassell education.
QA11.077 1992
510′.71—dc20 91–39854
CIP

ISBN 0–304–32553–8 (hardback)
0–304–32555–4 (paperback)

Typeset by Fakenham Photosetting Ltd, Fakenham, Norfolk
Printed and bound in Great Britain by Dotesios Ltd, Trowbridge, Wilts.

Contents

Preface to the First Edition

Mathematical education has become an established subject, related to both mathematics and education but deserving of specialized study in its own right. One important branch of mathematical education concerns learning, which clearly has close links with educational psychology: this book is about learning mathematics. There is now a very considerable body of knowledge, built up from countless research studies, which provides the student of mathematical education with a great deal of source material on aspects of learning mathematics, but introductory overviews are not common. This book is intended to provide an introduction to the study of learning mathematics that raises issues, provides discussion of theory and practice at a relatively elementary level, and supplies references for more detailed and specific follow-up reading. It is based on courses for higher degree and diploma students whom I have taught over a period of some fifteen years, and I am indebted to the many students who have both raised the questions and debated the issues with me. One problem in preparing the book was that it could easily have become too much like a catalogue of references, and many relevant references had, in the end, to be omitted. I hope that the many researchers who have provided invaluable knowledge but who have not been directly referenced in the book will not be offended. It is hoped that the book will provide an introductory text for all courses which include studies of learning mathematics.

Tony Orton
Leeds, 1986

Preface to the Second Edition

Since the first edition of this book was published, research and debate concerning children's learning of mathematics has naturally continued and knowledge has grown. This new knowledge is reflected in the amendments to the book. In particular, the volume of literature on constructivism has expanded to such an extent that a separate section has been included in this edition. This section aims to build on the references to constructing knowledge and understanding in the original text, but also to extend the ideas through discussion of more recent literature. In addition, the success of the book around the world has justified the inclusion of rather more international and cross-cultural references, though the main focus remains England and Wales, where the introduction of a National Curriculum has resulted in some changes to the original text. In fact, it can be argued that cross-cultural discussion is likely to enhance appreciation and understanding of issues and problems which themselves know no national boundaries. The interest of many teachers around the world in word problems and pupils' difficulties has led to a further new section. Minor amendments and additions have also been made in order to bring the book up to date.

Tony Orton
Leeds, 1991

To Jean
who helped.

Chapter 1

Do Teachers of Mathematics Need Theories?

Educational issues are rarely clear cut. An individual teacher may hold very firm views on a particular issue in mathematical education, but must at the same time accept that very different, even completely contrary, views may be held by a colleague in the same school. Examples are not hard to find. In recent years the availability of pocket calculators has sparked off discussion and controversy about how and when calculators should be used. If young children are allowed to use them will they ever learn their multiplication tables? Could sensible use of calculators enhance understanding? A variety of different kinds of structural apparatus exists for helping children to acquire the concepts of elementary number. Is such apparatus essential? Which is the best? Some teachers believe that mathematics should be a silent activity with each of the children always producing their 'own work', but other teachers value discussion between pupils. Is discussion important for all or do some pupils opt out and so learn nothing? Deciding on appropriate mathematics for older, low-attaining pupils has always been a problem: is social arithmetic the best answer or is it seen as irrelevant by the pupils? The debate about the place of calculus has continued throughout much of this century. Is there a place for calculus before the sixth form or is it conceptually too difficult for all but a very few? These are only a small selection of the many issues which would lead to debate and disagreement.

In accepting a particular viewpoint, or in taking sides on a particular issue, it could be said that a teacher has accepted a theoretical position. Throughout any day in school we are adopting particular ploys and using particular methods because we believe they work. Such limited theories are based on experience, intuition and perhaps even on wishful thinking. They may be helpful, they may, on the other hand, be dangerous. For example, is it dangerous to introduce division of fractions in the primary school? It might be if, in not understanding, children become frustrated and anxious and come to reject mathematics as a meaningful and worthwhile activity. It appears that the job of teaching cannot be done without accepting theoretical views, though such theories, it would undoubtedly be claimed, should be based firmly on empirical evidence. In this sense it appears that we do need theories as a basis for decision-making in the classroom.

However, although teachers do need to adopt and practise theories in their daily work it is not unusual to find many who are sceptical or even disparaging about the value of large-scale theories. Major theories which might enlighten the teaching–learning process are dismissed as irrelevant. It is possible that such theories are rejected without being given serious consideration. For example, it might not be appreciated by those of us who use and value structural and other apparatus that the invention of the equipment might have been prompted by acceptance of a particular learning theory. One of the earlier kits was devised by Stern (1953) because her belief in Gestalt theory demanded that such apparatus was available for children. Of course, it is possible that some of us reject theories because accepting them might necessitate adopting a radically different teaching style.

A theory should be based on observation of children's behaviour in learning situations. Subsequently, the theory should enable us to explain what we see in school and also to take appropriate action. In this sense our theory explains, and could even predict, phenomena. Hopefully, with sufficient data on which to construct hypotheses, our theory might present a systematic view of phenomena whilst at the same time remaining relatively simple to grasp. The large-scale, general theories which are sometimes rejected by teachers have usually been based on a systematic view extrapolated from a much wider range of events and situations than any one individual can have experienced and contemplated. The view which underlies this book is that education is too important for us to be able to dismiss as irrelevant theories of learning which attempt to do what has just been described. Child (1986) explained it by saying, 'innovation and speculation in learning . . . are more likely to succeed when they are informed by sound theoretical frameworks'.

One major problem is that there can appear to be a large number of conflicting or contradictory general theories in existence. Historically, two major kinds of theory have been developed, referred to here as 'behaviourist' and 'cognitive', and these two certainly do conflict, though recent work has attempted to reconcile some aspects. Within these two very different schools of thought there have been variations and amendments throughout this century. It is perhaps more important to think first about the major differences between the two and not to worry about differences within or in any overlap which might be thought to exist between them. The major differences can be explained by referring to a situation in learning mathematics.

It is very important that children come to an adequate understanding of place-value. At a certain stage in the education of young children it would be reasonable to ask them to write 'four hundred and twenty-seven' as a number. Some children would write

40027,
others 4027,
or even 400207,

and these would not be the only answers offered from within the class. Most children, it is hoped, would correctly write

427,

but the incorrect responses, however few, would require remediation. How should remedial action be taken? How should the children be taught the concepts in the first place?

If our theoretical view is that children learn through practising to produce the correct response to a given stimulus, then we should give them more practice. Such an approach might incorporate the use of apparatus, but the fundamental intention is to give practice. In this approach there might well be the underlying assumption that we are there to feed information and knowledge into the mind of the child. In an extreme form the approach might be referred to as rote learning.

If, however, we believe that children learn through making sense of the world themselves, we would wish them to discover the essential relationships through interaction with an appropriate environment. Thus we might well provide structural and other apparatus and devise activities and experiences, allowing exploration of the structure of the situation. It would, of course, be necessary to ensure that the notation emerges as being logical and efficient, so some teacher intervention is inevitable. In this way understanding would grow from within, as it were. Any attempt to hasten the child by injecting rote methods might not only be unsuccessful, it might persuade the child that mathematics is meaningless and worthy only of rejection.

It should be stressed that these two contrasting approaches are not intended to explain fully the difference between particular behaviourist and cognitive approaches, they are merely intended to illustrate how possible interpretations might manifest themselves in mathematics lessons. It would be wrong to tie rote learning too closely to the behaviourist approach and by implication suggest that it has no place within a cognitive approach. There is, after all, the eclectic view, that children do need to develop their own understanding from within, but that there might be a very firm place for practice, and even perhaps for some element of rote learning.

It is unfortunate that conflicting theories and variations on a common theme might lead some teachers to reject them all. Some conflict is, after all, only to be expected within a discipline with a very short history. It is sometimes forgotten that the so-called 'pure' sciences have been the subject of many battles over many hundreds of years. Even now, disagreements can still exist. Scientific theories are continually being modified, elaborated and simplified and, from time to time, radically new ideas are produced. In the world at large decisions have to be made, and they are made on the basis of existing theoretical views. Not all such decisions ultimately turn out to have been correct. Particular theories of learning might also be wrong, or might need qualification or amendment. But the formulation of a theory and the observation of it in action are both part of the process through which we improve our understanding. We can learn more about the learning process if we are prepared to encourage the formulation of theories and then test out those which appear most likely to help.

Learning is a mental activity. We might therefore understand more about learning if we knew more about the functioning of the brain as a processor of information. The brain receives information, interprets it, stores it, transforms it, associates it with other information to create new information and allows information to be recalled. In recent years considerable attention has been accorded to the information-processing aspect of learning theories, and this has led to considerable interest in what goes on inside the brain. It has been known for many years that different learning activities take place in different parts of the brain, though that very simple statement unfortunately glosses over complexities which are certainly beyond our scope for the present.

The relationship between the chemistry of the brain, the nerve impulses which are generated and learning are, likewise, too complex for this book. It must be clear, however, that we might understand much more about learning, as an aspect of psychology, when we understand more about the workings of the brain as an aspect of physiology.

One of the traditional justifications for teaching mathematics is that it teaches logical thinking. Unfortunately, the logic of mathematics is not necessarily the same as the logic of any other sphere of human intellectual activity. The argument therefore stands or falls on the theory that the ability to think logically in mathematics is a transferable skill and can be put into practice outside mathematics. This assumption has been known in the past as 'transfer of training'. Shulman (1970) said, 'Transfer of training is the most important single concept in any educationally relevant theory of learning.' There is no doubt that the former view that studying geometry or Latin made one a better logical thinker is now completely discredited. Nevertheless, some lateral transfer must be possible, lateral implying the transference of skill in one domain to the achievement of a parallel skill in another domain (though 'parallel' is not easy to define in this context), for, without it, learning would be extremely slow and would be limited to what had actually been encountered in the course of instruction.

There is no general agreement about the extent to which lateral transfer can take place in mathematics. There have been psychologists and learning theorists who have expressed the view that broad transfer can take place, that ideas and strategies can be transferred within a discipline and perhaps even outside. Thus it might be believed that mastery of the idea of balance, as a physical property using weigh-scales and weights, can be transferred and applied to the solution of linear equations, and might even be transferable to studies of balance in nature and balance in economics. It might also be believed that learning how to prove results in Euclidean or any other sort of geometry would be transferable to proof in other branches of mathematics, to proof in other disciplines such as science and even to proof in a court of law. Other psychologists have believed that transfer only occurs to a very limited extent, perhaps only to the extent that identical elements occur. This latter view probably carries more conviction than the former at the present time. Some transfer must be possible, but it will probably be limited and might depend on the conditions under which learning takes place. It is certainly not wise to assume that transfer of skills will occur when teaching mathematics.

The learning difficulties which one observes as a teacher of mathematics raise many other questions for which one might seek an answer from theories. For example, although reflection on our own experience should suggest to us that learning cannot be achieved in a hurry, some children appear to learn incredibly slowly. What determines the rate of learning? Some children make very rapid progress, a few even make astounding progress given the opportunity to learn at their rate rather than the class rate. Is it possible to accelerate the learning of mathematics for more pupils or even for the majority of pupils and, if so, how? At the moment it seems that for many children it is not a matter of whether they can learn mathematics more quickly; rather it is a question of why they appear to take in hardly anything at all. Is it that mathematical ability is a peculiar aptitude possessed by only a few?

Individual differences are very significant in many spheres of human activity. Some of us are barred from particular occupations because of physical characteristics, like being too small, too overweight, or having poor eyesight. Many of us who have become teachers of mathematics because of an apparent aptitude and a liking for the subject would not have been able to become teachers of other subjects, like English or history. Amongst international athletes some are good only at running, others at jumping events and yet others at throwing events. Individual differences might be important even within mathematics. Hadamard (1945), in discussing mathematicians, drew attention to great differences in the kind of mathematical aptitude which individuals have displayed. In the classroom it might be that different learning environments and different teaching styles are needed for different pupils, which would present very great teaching problems in the sense that any individual teacher also presumably has preferences which are in accord with only a proportion of the pupils. Any acceptable theory which enables us to understand individual differences would be very valuable.

One interpretation of the evidence of what children appear to learn and appear to have difficulty with is that there are serious stumbling-blocks in the logical structure of mathematics. With many young children, the ideas of place-value appear to present hurdles which cause frequent falls. With slightly older children the introduction of algebraic notions causes problems for which some pupils, in later life, never forgive us. There are mathematical ideas, like ratio and rate, which continue to cause difficulty for many throughout adult life, even though they are extremely important ideas. It is possible to survive in life without understanding the implications of a fall in the rate of inflation, but it is a pity that so many adults have learning difficulties with mathematics. So what is it about particular aspects of mathematics, algebra and rate of change, which makes them so difficult? When we analyse the structure of mathematics in order to devise the optimum teaching sequence, how do we allow for the fact that the logical order of topics might fail us for psychological reasons?

A major complexity in learning any subject is the relationship with language learning. At a surface level the effects may be observed when a child cannot do the mathematics because the particular language used is not understood. There are many examples of peculiar language and of familiar words used in different or very specific ways in mathematics. At a deeper level, to understand the language is to understand the concept which a particular word symbolizes. More fundamental still is the relationship between language and learning. Does language merely enable one to communicate learning that has already taken place? Is language the vehicle which enables us to formulate our ideas and manipulate them to create new meanings? Is it that language development is inextricably tied to overall cognitive development and cannot be thought of as a separate entity?

It has been suggested earlier that the learning environment might be an important factor in promoting the understanding of mathematics. It might, therefore, be postulated that the richer the environment the more efficient the learning, but to some extent that begs the question. What constitutes a rich learning environment in a subject which is basically a creation of the human mind and in which the aim is to enable abstract argument to take place through the manipulation of symbols? The belief that young children must be allowed and encouraged to interact in a very active manner with physical or concrete materials is a theoretical stance suggested through experience of teaching young children (though not all young children are provided with such an environment). If we accept this and provide an environment rich in equipment and learning materials for young children, how soon can we wean them from it? Do we need to do anything for older children in, for example, coming to terms with algebra? Or should we not be attempting algebra until the pupils can manage without concrete apparatus? When can children begin to learn only from exposition and from books?

These are some of the many aspects of mathematics learning for which we might seek answers. Many of the theoretical viewpoints expressed in subsequent parts of this book do attempt to account for questions raised above. It has already been suggested that teachers need theories, hence major theories of all kinds and from many sources are included within the discussion of particular questions. First, however, we take a look at the problems from the child's point of view, to see what is being learned and how thoroughly. We investigate the empirical evidence on which, it has been declared, theories might be formulated.

QUESTIONS FOR DISCUSSION

1. What are your current beliefs about the most effective ways of promoting the learning of mathematics? (These beliefs might change as a result of reading this book!)
2. To what extent should the teaching of mathematics be intuitive and pragmatic and to what extent should teachers be deliberately trying to put theories into practice?
3. What issues concerning mathematics-learning and their implications for teaching are debated or discussed in your school? What should be discussed?
4. What mathematical topics or ideas which seem to fit logically into your ideal teaching scheme for given groups of pupils appear to be particularly inappropriate because pupils experience learning difficulties?

Chapter 2

What Mathematics Can Children Learn?

INVESTIGATING MATHEMATICAL UNDERSTANDING

As teachers, we may be involved in writing syllabuses and preparing detailed schemes of work. Many of us are guided, or constrained, in such tasks by a National Curriculum. In this planning, and indeed in the preparation of a National Curriculum itself, it would seem to be important to take into account the evidence of what children appear to be able to learn. There is no point in defining unreasonable objectives, but we do need to extend the knowledge and understanding which our pupils have. We may also need to take note of suggestions that pupils in countries other than our own are achieving more. But in seeking the right middle road for all our pupils we often get it wrong for many. On the one hand, evidence of children failing to learn because of unreasonable aims has been ignored in our enthusiasm as mathematicians to impart as much of our own mathematical knowledge as we can in as short a time as possible. On the other hand, there is a strong feeling that some groups of pupils are not being sufficiently extended—though this is likely to be a minority problem.

At most stages in the education of the majority we find an overloaded mathematics curriculum, with pupils hastened along through material which, at best, is only half learned. It must be admitted that in no way can we achieve complete mastery of anything, in the sense that there are always possible extensions. The point at issue is whether the pupils achieve adequate mastery to enable them to proceed with what we have decided comes next. The evidence we have suggests that all too many pupils frequently fail to match up to our expectations in this respect. We, the teachers, can be very easily misled. Young children can learn to recite numbers long before they fully comprehend what the numbers represent and how they are related, and we can easily assume they know more than they do. They can appear to be attaining correct answers to what we set, but they might be slavishly following the routine we have suggested and might not grasp why it works. Her Majesty's Inspectorate (HMI) (1985) have suggested that 'some standard written methods of calculation, such as a long division, which many pupils find difficult and few really understand, should no longer be generally taught'. Yet it could be said that it is the widespread availability of cheap calculators which has prompted this view, not the difficulty of long division. We certainly cannot claim that we have not known that long division was difficult. Renwick (1935), for example, reported American research that 'the optimum mental age for beginning to learn long division is 12 years 7 months'. Of course, although a topic like long division might come to be regarded by teachers as less vital than it once was, and too difficult anyway, this view might not be readily accepted by others in our community, such as parents.

Long division is not the only topic in our packed primary school mathematics curriculum which pupils find difficult, and which perhaps should be deferred (if taught at all). The primary curriculum convincingly illustrates how our enthusiasm to introduce pupils to all the interesting mathematics we can think of blinds us to the magnitude of what faces the average child. Of course we wish to extend our pupils. Of course we wish to provide for the most able. But freedom from restrictions created by national or regional selection examinations at eleven plus led to a considerable broadening into more interesting topics without corresponding reduction of less vital ones. Why does work with fractions still constitute such a substantial part of the primary curriculum? Nearly all of what we teach of fractions in the primary school is re-taught to most pupils in the secondary school because they have not achieved mastery. Although we may pride ourselves that we can explain to our own satisfaction why the method of division of fractions works, very few pupils grasp our explanation. We can often mislead ourselves that, because our explanation has been lucid, clear and logical, the message has been received. Sadly, this is not the case, but we should not be surprised. After all, many shots at goal miss the target! In the case of fractions, why is it necessary that children should be able to add, subtract, multiply and divide a wide range of sometimes awkward pairs? Essential ideas illustrated by very simple cases might make much more sense, particularly at primary school level. It was Skemp (1964) who, in devising his own curriculum, stated the view that fractions provide the obvious example of a mathematical idea previously considered to be elementary which analysis of concepts reveals as far from simple. The message itself was not new and was comprehensively discussed in Renwick (1935).

The same problem has arisen in the secondary school, where algebra has caused a very serious learning problem. One can sometimes find that textbooks with a high algebraic content and intended for brighter pupils are in use with a very wide range of pupils. Despite earnest declarations when the Certificate of Secondary Education (CSE) was introduced that it must not become a watered-down General Certificate of Education (GCE), that is exactly what it did become for many pupils. Cockcroft (1982) was very critical on this point, suggesting that the majority of secondary school pupils were following syllabuses which were of a difficulty and extent appropriate only to about a quarter of pupils. Again, this message was not new. It remains to be seen what happens as new schemes of examining begin to exert their influence on our curriculum and whether we can really determine the right curriculum first and let assessment follow. Recent indications from syllabuses of the General Certificate of Secondary Education (GCSE), which has replaced both Ordinary level GCE and CSE, together with the National Curriculum of England and Wales, suggest that the inherent difficulties of algebra, as regards the majority of pupils, have been to some extent acknowledged. Overall, however, it seems that we, as teachers of mathematics, have a very poor record when it comes to taking into account evidence of what children can learn.

Of course, it is not easy to be sure of what children can learn. We may feel that we have imperfect, even conflicting, evidence on which to base our decisions. In the first place we do not know what children *can* learn, only what they appear to have learned. We are forced to use measuring instruments which may be far from perfect. Certainly the evidence gained from typical school tests and written examinations may be very flawed. Probably the best vehicle for investigating what has really been learned and what misunderstandings and misconceptions still remain is the individual interview, used by Jean Piaget over many years and by countless other researchers as they discovered its value. In this method the teacher, or researcher, asks the child questions, records the responses, and subsequently analyses the interview data obtained. That sounds very simple, and is certainly possible, occasionally, in a normal classroom situation. There is, however, normally rather more to it than this simplistic outline suggests. Questions or tasks need to be carefully structured, sequenced and standardized, so that there will be validity to data collected from a large number of interviews with different children. It is often necessary to have alternative, but also standardized, questions, and to be ready with appropriate supplementary questions to be used according to the initial response or reaction of the child. Most research of this kind has been carried out by teachers, either alongside their normal teaching or in association with a research project to which they might even have been seconded. It is not always easy to train oneself to become a researcher of this kind. It is very easy to fall back into the role of the ordinary teacher and make suggestions, to talk too much instead of listening. It is very hard to allow free rein to the child whilst still being systematic in the collection of evidence. Piaget (1973) summed up the problem as follows: 'The good experimenter . . . must know how to observe . . . [but] must constantly be alert for something definitive, at every moment he must have some working hypothesis, some theory, true or false, which he is seeking to check.' In the light of Chapter 1 of this book, it is interesting to note the reference to theory in Piaget's statement. Of course, we do not need to collect data and leave the child in ignorance. We can subsequently try to remedy the weaknesses of our children all the better for having an accurate picture of what those weaknesses are.

Even if one is able to collect useful research data there can be other problems, for example, apparent inconsistency. Sometimes it seems that children can answer a question on one day but not on the next. Sometimes it seems that children can answer one question but cannot answer another which seems to us to be exactly alike. Sometimes our evidence might suggest that task A is harder than task B, and sometimes the reverse might appear to be true. Sometimes the accessibility of appropriate language might not be adequate for the children to convey a true picture of what they really do understand, but how do we know? Sometimes the variety of responses, the complexity of the data collected, almost defies analysis. Often, children are learning as we question them. Such are the difficulties, but all research must present comparable difficulties. Despite these problems, and possibly many others, the evidence strongly suggests that there are widespread misconceptions, that there are limits in terms of levels of understanding which are achievable, in terms of rate of progress and of many other facets of learning mathematics.

We must also take into account that we all have a greater capacity for learning when we really want to learn. We cannot ignore the effect on quality of learning of motivation, interest, determination and the desire to succeed. The search is continually on to find ways of making school mathematics more appealing and exciting, more relevant and useful, because we know that children learn much better under such circumstances. High levels of motivation and interest in individual children who do not shine in school mathematics can make them experts in very diverse areas of knowledge from pop records to computer programming, from football to British butterflies. At the opposite extreme is the problem of de-motivation through anxiety. This might be created through unsuitable subject matter, unsympathetic teaching and a whole variety of environmental factors. Some children do appear to panic quite badly, and this is clearly not helpful in fostering learning (see Buxton, 1981). We do not know what such anxious children might achieve under different circumstances. It is not possible to completely separate out the cognitive factors from the affective (where 'cognitive' may be thought of as pertaining to the recall or recognition of knowledge and the development of intellectual abilities and skills and 'affective' as pertaining to interest, attitudes, values and appreciations—see Bloom *et al.*, 1956).

It could also be said that learning might be enhanced if all teachers were as good as the very best. At the same time, it would be very difficult to define good teaching. Certainly, in any attempt to improve the quality and quantity of learning it is the teacher who plays the most important part and not the teaching materials, the classroom or the syllabus. Nevertheless, the information that we have of children's cognitive development paints a detailed picture of how things actually stand at the moment. That is what we really want to know in order to seek improvement. It is also what needs to be explained by theories of learning. In order to elaborate on what children can learn empirical evidence obtained from pupils' performance in a number of topics across the complete 5–18 age range is now presented. It should be pointed out that what follows is a very limited selection of illustrations when one considers the amount of such data now available across all branches of mathematics.

PLACE-VALUE

Our modern number system, based on symbols for the digits and including a symbol for zero, took mankind a very long time to develop. With these ten symbols we can represent numbers using place-value, and so this notion is a major early idea which children need to master, before, for example, they can progress with confidence through the four operations on numbers, namely addition, subtraction, multiplication and division. Considering that our present number system did take a long time to develop it should not be surprising that some children are very slow to grasp the full implications of the notation and its underlying conceptual structure. See Branford (1921) for an early discussion of this issue. It is important to consider place-value in the context of this chapter because of its importance as a major conceptual basis for all number work.

One particular experiment on children's understanding of place-value was carried out by Barker (1979) who tested and interviewed a very large number of children in the 7–9 age range in a variety of different schools. Some of his tasks and the corresponding results are presented below. The first of these tasks was presented orally, and the other three were presented in both oral and written forms.

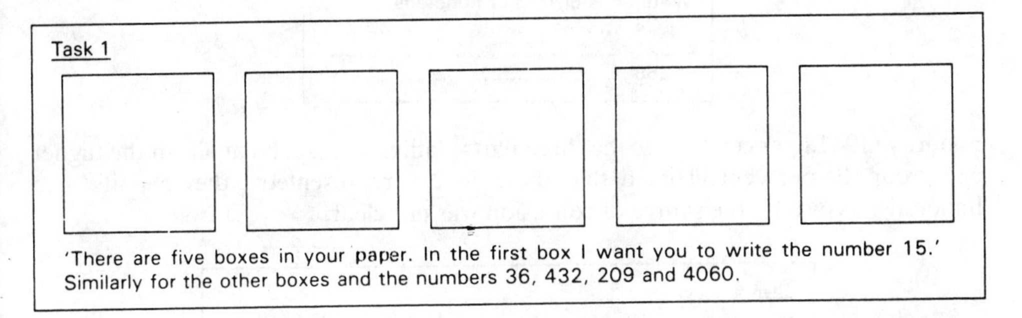

Most children were able to write 15 and 36, though three children reversed the digits in 15. There were 11 different responses in writing 432, only seven of which were understandable. This illustrates two of the major problems of analysing research results: the variety of responses and determining what thinking produced them. In the 7–8 year age group as many as 27 per cent wrote either 4032 or 40032. For 209, 25 per cent of this same group wrote 2009, but the percentage was only five in the 8–9 age group. Such an improvement between two consecutive year groups is, unfortunately, not common in mathematics-learning particularly when one looks at results concerning much older pupils. There were 21 variations for 4060, all but one containing the correct digits. The errors 40060 and 400060 were common in both age groups with 50 per cent of 7–8-year-olds and 10 per cent of 8–9-year-olds giving such responses. Only 17 per cent of the 7–8 year group, but 78 per cent of the 8–9 year group, gave the correct answer.

Task 2

In a skipping game 5 children got these scores:

Peter 25, Mark 125, Tony 19, Karen 152, Lisa 78.

Who got the biggest score? ....

Who got the smallest score? ....

One particular error predominated in the first part of Task 2, namely choosing 125 instead of 152, though most of the errors came from the lower age group. Again, the difficulties of making sense of what children say are illustrated by one boy who said that 125 was bigger than 152 because they both have one hundred but 125 has a twenty and 152 has a 15, and by another boy who claimed that 125 and 152 were both the same. In the second part of the question 15 per cent of 7–8-year-olds and 19 per cent of 8–9-year-olds chose 25 to be smaller than 19. However, when interviewed subsequently with the same numbers in a different order, all these children chose the correct smallest number. The fact that 25 was the first number in the list might have been significant. Other children selected other wrong answers, in fact the correct answer to the second part was chosen by only 65 per cent of the lower age group and 78 per cent of the higher.

Task 3 proved difficult, especially for the younger age group, with a significant

Task 3
What does 5 mean in these numbers?
Write units or tens or hundreds.
35 ....
521 ....
256 ....

minority (10–18 per cent across the three parts) failing to respond at all. In the higher age group 19 per cent claimed that the 5 in 256 represented either 5 units or 5 hundreds. Typically, the source of confusion was not clear.

Task 4
(a) 'There are three numbers, 2, 8 and 9. I want you to arrange them to make the biggest number you can. You have to use each number once and use all three numbers.'
(b) Similarly for the smallest number.

There were 32 variations for the biggest number from the three digits, 13 of which included one or more zeros, presumably to make the number bigger. This suggests that many children were struggling to master the real meaning and significance of zero. In fact, a small number of children gave the answer 19 (the sum of the digits) and a further two children gave 91. For the smallest number there were 23 variations, only four of which incorporated zeros.

It would be a very serious mistake to assume, because the older age group were so much more proficient than the younger, that mastery of place-value in all its ramifications is eventually achieved by the majority of children. Clearly, such ideas as those above are eventually mastered by many children, though some might struggle for a time, but there is ample evidence of difficulties in later years. A different experiment, organized by the present author, was based on a group test with 829 pupils across the 11–16 age range and was carried out by a group of teachers. The question sheets presented to pupils contained only four sets of numbers. For the rest, it was up to the teachers to use the identical forms of words:

> These are some number puzzles. There is at least one answer to each question and there may be more than one.

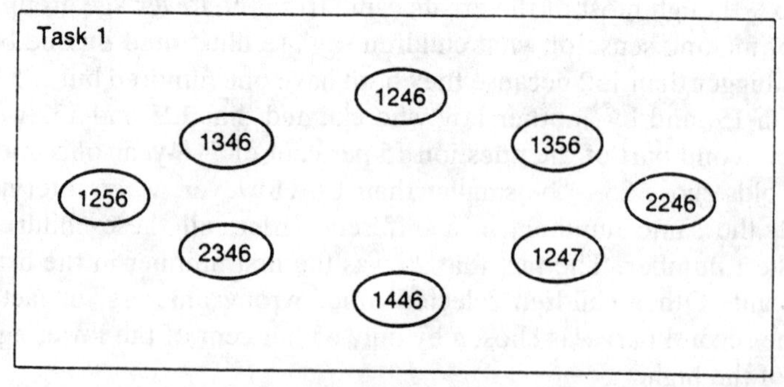

In the first question I want you to link any pairs of numbers in which one number is 10

more than the other. For example, if you had 7 and 17 there in the ring of numbers, you would link those because one is 10 more than the other.

Show on blackboard:

Now write '10 more' by the question to remind you, and then go ahead.

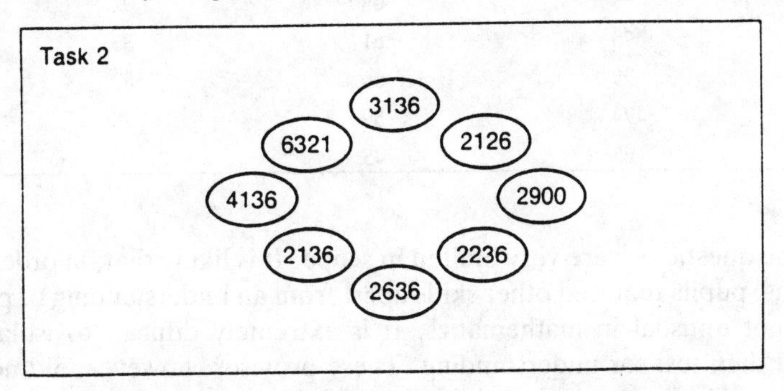

In the second question I want you to link any pairs of numbers in which one number is 100 more than the other. For example, if you had 7 and 107 there in the ring of numbers, you would link those because one is 100 more than the other.

Show on the blackboard:

Now write '100 more' by the question to remind you, and then go ahead.

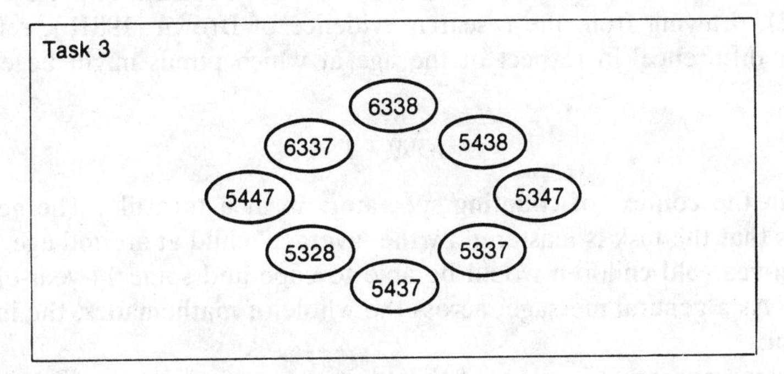

In the third question I want you to link any pairs of numbers in which one number is 1000 more than the other. Write '1000 more' by the question to remind you, and then go ahead.

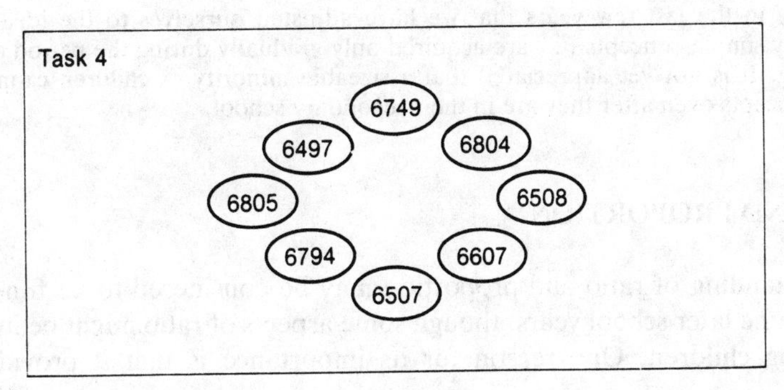

In the fourth question I want you to link any pairs of numbers in which one number is 10 more than the other. Write '10 more' by the question to remind you, and then go ahead.

Table 1 Percentage of correct responses in each age group.

	1	2	3	4
15–16	88	68	92	61
14–15	75	61	83	72
13–14	79	66	82	56
12–13	74	59	79	54
11–12	65	45	72	30

Clearly the questions were very limited in scope. It is likely that, in order to answer the questions, pupils required other skills apart from an understanding of place-value, but this is not unusual in mathematics. It is extremely difficult to isolate a single concept and then test for understanding. These provisos, however, did not alter the fact that the pupils did not do as well on the questions as the teachers expected. Low-ability children, who by definition one would expect to fare badly, found the test even more difficult than was ever imagined, and other pupils demonstrated considerable weaknesses too. These are shaky foundations on which to build.

The concept of place-value and how well or otherwise children understand the notions has been discussed in more detail elsewhere—see for example, Brown (1981b) and Dickson, Brown and Gibson (1984). In the context of place-value Cockcroft (1982), drawing from the research evidence of Brown (1981b), referred to a 'seven-year difference' in respect of the age at which pupils might be expected to answer

$$6399 + 1$$

correctly, in the context of counting spectators in at a turnstile. The general conclusion was that the task is mastered by the 'average' child at around age 11 but that some seven-year-old children would be able to cope and some 14-year-old children would not. As a general message, across the whole of mathematics, the implications are enormous.

As suggested earlier, it is not as if this kind of knowledge of pupils' difficulties is only just becoming known. But it might be that not enough people have been aware of their extent. Fogelman (1970) wrote:

> It is only in the last few years that we have adjusted ourselves to the idea that . . . seemingly simple concepts . . . are acquired only gradually during the period of primary schooling. It is not yet appreciated that a sizeable minority of children cannot handle these concepts even after they are in their secondary school.

RATIO AND PROPORTION

An understanding of ratio and proportion may be considered to be fundamental to learning in the later school years, though some aspects of ratio might be introduced to quite young children. One reason for its importance is that it provides a useful problem-solving technique. Given that seven pencils cost 63 pence, the cost of three pencils could be obtained by using the ratio $\frac{3}{7}$, and the solution obtained from the proportion

$$\frac{3}{7} = \frac{}{63}\ .$$

Beyond this elementary application, other mathematical concepts build on the basic ideas. Geometrical theorems and results based on similarity and on parallel lines and intercepts require an appreciation of proportionality. The idea of gradient, which is important in the algebra of graphs and in calculus, also depends on ratio and proportion. Simple trigonometry, likewise, has its beginnings in a study of equal ratios. Rational numbers are studied throughout most years of a child's school life, progressing through operations on fractions, decimals and percentages, and perhaps culminating in a more formal study of the number system itself. Ratio also underlies pie charts, scale factor and the slide rule. It is clear that it is important to discuss proportionality in the context of this chapter because it pervades mathematics.

The development of scientific understanding also relies on the ability to handle ratios, for example, in the definitions of density, velocity and acceleration; in calculating chemical equivalents; in applications of the ideal gas laws and in using many laws of physics. Other school subjects make use of proportionality through simple calculations such as the pencils problem given earlier, through percentages, through scale and through graphical representation. Ratio and proportion are important in mathematics, science and elsewhere but pupils often struggle and find it difficult to grasp the use of proportionality.

Some insight into pupils' reactions to proportionality problems can be gained by considering typical responses. A widely used task for testing understanding concerns measurement and calculation based on the heights of two pin men. In the task, first used by Karplus and Peterson (1970), a diagram of Mr Short and Mr Tall is made available to pupils (see Figure 2.1), and two different measuring techniques are described, using objects such as paper clips and buttons. Thus Mr Short is measured first, using paper clips, and the pupil finds that the height is four clips. Mr Tall is measured next and he is found to be six paper clips in height. Finally, Mr Short is measured with buttons, and he is six buttons in height. From the tabulated results the pupil is asked to state how big Mr Tall would be if measured with buttons, and then to state a reason for the answer.

	Height	
	Paper clips	Buttons
Mr Short	4	6
Mr Tall	6	?

Some typical responses are as follows:

'Eight, because Mr Tall is two bigger.'
'The paper clips are about twice the buttons, so Mr Tall is $6 \times 2 = 12$.'
'Mr Tall is two more than Mr Short and the buttons are twice, so Mr Tall is $6 + 2 \times 2 = 10$.'
'Mr Tall is half as big again, so he is 9.'

In any year group the percentage of correct responses from a sample of pupils

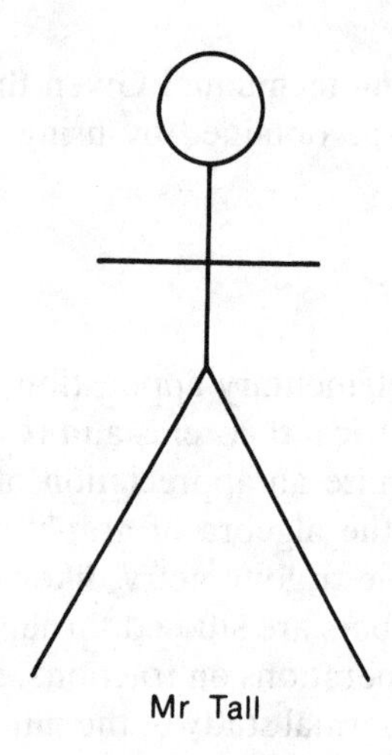

Figure 2.1

across the ability range can be surprisingly small. More able pupils naturally have a higher success rate than other pupils, and average and less able pupils, the majority of a year group, resort to incorrect methods of solution. Probably the most common incorrect response is some variation on an additive approach, as in the first response quoted. In an experiment by the author (Orton, 1977) carried out with a group of teachers with 186 pupils in the 11–16 age range, only around 20 per cent of responses showed correct use of proportion whereas around 50 per cent used addition or some closely derived methods. Incorrect methods were not confined to younger pupils either. A large number of responses appeared to have been arrived at by guesswork.

Research results also confirm that the complexity of the proportionality rule makes a big difference. The following three tables for 'find the missing number' puzzles reflect increasing complexity from 1 : 3 through 2 : 3 to 4 : 3, though they also incorporate a difference in design between (*a*) and the other two. These puzzles were used with the same 186 pupils referred to previously, the main objective being to reveal to their teachers the extent of pupils' difficulties with ratio.

(a)			*(b)*			*(c)*		
	1	3		2	3		8	6
	2	6			9		10	$7\frac{1}{2}$
	6			8	12			9
	8	24		14	21		28	21

Although there were similarities between the response patterns for (*b*) and (*c*), there was no doubt that pupils found (*c*) more difficult than (*b*). Part (*a*) was found to be comparatively simple and, for able pupils, a much higher success rate was reported on (*a*) than on the Mr Short and Mr Tall problem. For all abilities there were fewer additive attempts on (*a*) than with Mr Short and Mr Tall. Other research, for example, that reported in Hart (1981), confirms the comparative simplicity of ratios of the form 1 : 2 and 1 : 3, particularly in comparison with ratios of the form m : n, where neither m nor n equals 1.

Although additive responses were not so frequent with the number puzzles, it is interesting to note that, even when answers were correct, some pupils showed that they were still inclined to use some element of addition. In another experiment (Orton, 1970), involving 72 pupils, responses to task (*c*) were analysed as regards *type* of correct explanation. Of the responses 34 were correct and 25 used only proportion, for example: 'The second number is three-quarters of the first.' The other nine pupils variously said: 'Take one-third of the second number and add it on to get the first', or, 'Because one quarter of the first number is taken away each time', or similar. Just one response was, in effect, 'From six to seven-and-a-half to nine you add one-and-a-half on each time; from eight to ten to twelve you add two on each time'. It is worth noting that, when pupils give a correct answer, this does not imply that their thinking corresponds to that of the teacher; indeed they may not have used the expected reasoning at all.

It is clear that there are many mathematical topics which involve ratios. As already mentioned, fractions may be much more difficult to handle than we have imagined in the past. The underlying method of using equivalence, which has been in use in recent years to bring some unification to operations with fractions, implies using proportions. Thus, for example,

$$\frac{2}{3}+\frac{1}{4}$$

has to be rewritten as

$$\frac{8}{12}+\frac{3}{12}$$

by making use of the equivalences

$$\frac{2}{3}=\frac{4}{6}=\frac{6}{9}=\frac{8}{12}=\ldots\ldots$$

and

$$\frac{1}{4}=\frac{2}{8}=\frac{3}{12}=\frac{4}{16}=\ldots\ldots$$

The method has only limited success with many pupils. Experience with average pupils suggests that they may well be able to cope but remain largely at the level of applying a learned routine.

A method of calculation which appears to have decreased in popularity in recent years is sometimes referred to as the 'unitary method' or as 'the method of practice'. This consists of the solution of problems involving proportion by reducing to a unit, thus, if 7 pencils cost 63 pence, then 1 pencil costs $\frac{63}{7}$ pence and 3 pencils cost $\frac{63}{7} \times 3$ pence. The more sophisticated treatment leaves all computation to the end, but obviously intermediate calculations can be carried out, and many pupils prefer the security of completing divisions and products when they arise. The unitary method is not without its problems for children, particularly in distinguishing between direct and inverse proportion. However, one might argue that the method is appropriate to children who cannot yet confidently handle a direct method involving proportion. Such children will have problems however we try to simplify the procedure, but having a simple procedure available might be better than leaving children completely at sea. In the case of the unitary method, as with the case of fractions, an extreme argument is that, since pupils seem unable to understand the mathematics in the way the teacher would really like them to understand, it would be better to defer all such mathematics until they can. The counter-argument is that this might produce school leavers for whom the chance to use algorithms in important areas of numerical work has not been afforded. This is the kind of difficult decision that teachers have to make, particularly those operating in a formal class-teaching environment, and they cannot in so doing ignore the reaction of society as a whole. A carefully devised programme of work geared towards a facility only in limited areas of numerical work might, after all, accelerate the growth of understanding of proportionality.

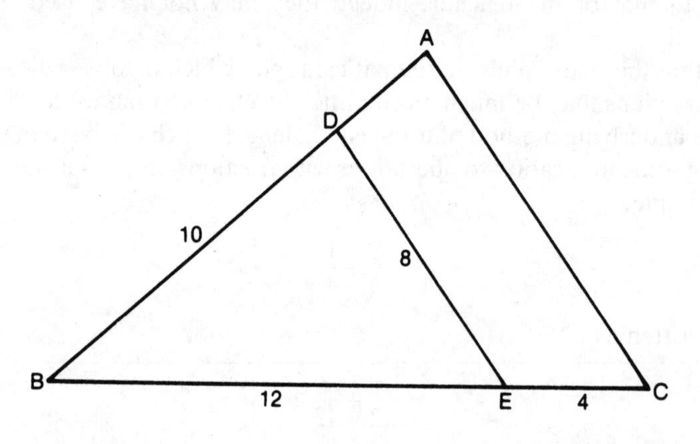

Figure 2.2

In geometrical terms the fundamental use of proportionality is through enlargement. Nowadays this is very likely to be approached through transformation geometry, whereas in the experience of many mathematics teachers it was approached through similarity, beginning with an extensive study of similar triangles. Whatever the beginnings, before long similar right-angled triangles will be used to introduce trigonometrical ratios. It might be thought that, when pupils have a diagram to illustrate the proportionality situation, as they do in geometry, they are better able to perform calculations, but there are still likely to be difficulties. The following short extract from a lesson illustrates the kind of problems that pupils can have. The class involved comprised able pupils of 12–13 years of age in a selective school.

Teacher (T): Here is a diagram on the blackboard (Figure 2.2). It consists of two triangles *ABC* and *DBE* and *DE* is parallel to *AC*. I want you to tell me the similar triangles so that we can calculate *DA*.
Pupil (P1): $DA = 4$.
T: Why?
P1: Well, *EC* is 4 and $DA = EC$.
T: Why?
P1: Because *DE* is parallel to *AC* and the distance between two parallel lines is always the same.

T: No, it is the perpendicular distance that is always the same. DA and EC are not perpendicular to the parallel lines. Can anyone tell me the similar triangles?
P2: Why don't we just measure DA?
T: We're not doing it by measuring, we are doing it by equal ratios.
P2: But it's a lot easier by measuring.
T: Which triangles are similar?
P3: Triangles ABC and DBE.
T: So how do we find DA?
P4: EC is 4, which is one-third of 12, so DA must be one-third of 10.
T: Look, what did we do about ratios of sides last time? Can anyone give me the equal ratios?

From this point on the treatment of this problem proceeded as the teacher had originally expected, from

$$\frac{AD}{CE} = \frac{DB}{EB}$$

to the answer.

The contribution of the first pupil illustrates a common geometrical confusion in connection with parallel lines. After the teacher's explanation, the expression on the pupil's face clearly suggested continuing confusion, though it should have helped him to sort out his thinking. The second pupil's contribution is also common, and indicates a preference for a concrete approach to the problem, an approach which in any case requires a scaled drawing. Pupil 4 contributed a wonderful, simple solution which was not accepted by the teacher because it did not correspond to the expected answer, again warning us that correct answers may be arrived at in a number of ways. Clearly, as teachers, we must never expect only one method to be used. Perhaps if all pupils could argue as Pupil 4 did, there would be no need for the orthodox method, which, in any case, for many pupils becomes a learned routine.

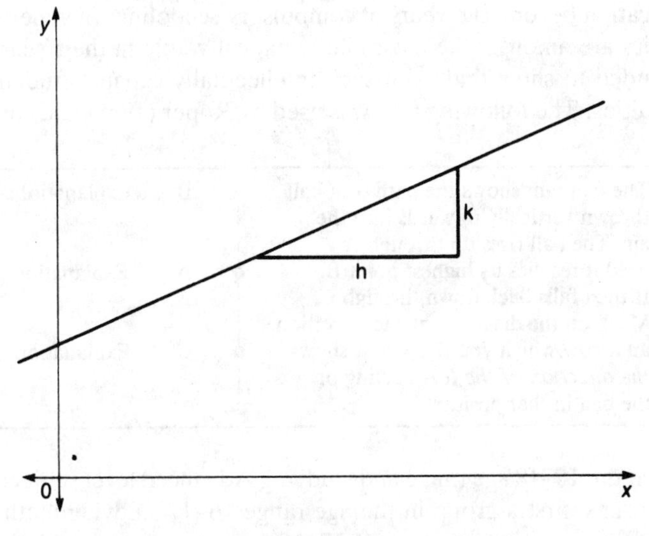

Figure 2.3

The approach to gradients of lines in graphical work is also based on the fact that ratios of corresponding sides of similar right-angled triangles are equal (see Figure 2.3). The technique of drawing any right-angled triangle and calculating

$$\frac{k}{h}$$

requires that the pupil does understand that you obtain the same ratio whatever the size of right-angled triangle, an obvious manifestation of proportionality. Teachers can emphasize the method of going one unit along from any point on the line in the x direction, and then going parallel to the y direction until the line is met again, and this may be all that can be achieved with many pupils, but it implies nothing about understanding the situation as the teacher understands it.

The occurrences of proportionality in school mathematics are many and varied. True understanding of proportionality develops very late, if it develops at all, in a child's school life. The difficulty of proportionality has been appreciated for a long time. Renwick (1955) said, 'This concept is shown to be far beyond the intellectual range of intelligent girls of eleven. . . .' For intellectually weaker children it is beyond their range at fourteen, and for some it may never be within their range. We must take this not as an indication that proportionality must be avoided at all costs in our teaching, but as information and guidance in helping us to understand the difficulties of the child when proportionality inevitably arises. It is possible that ratio and proportion could be taught in such a way that learning is optimized, though possibly not greatly accelerated. In this respect, and in respect of much more detailed information about pupils' understanding of ratio, important references are Hart (1981) and Hart (1984).

MECHANICS

Mechanics is learned, within mathematics, by only a proportion of those students who remain in education beyond the years of compulsory schooling. In general, those who study mechanics are amongst the most mathematically able in their year group. This section is included to show that even such intellectually capable students may have learning difficulties. The following task was used by Roper (1985) and concerns force.

The diagram shows the path of a ball thrown vertically upwards into the air. The ball rises up through A until it reaches its highest point B. It then falls back down through C. Mark on the diagram, at each position, an arrow which you think best shows *the direction of the force* acting on the ball in that position.

B Explanation:

A Explanation:

C Explanation:

The students, in the 16–18 age range and studying Advanced level subjects, were from four distinct groups: first a group in the age range 16–17 studying mathematics with mechanics (LM), secondly a group in the 17–18 age range studying mathematics with mechanics (UM), thirdly a group in the 17–18 age range studying physics but not mathematics (UP), and finally a group in the 16–18 age range not studying mechanics in either mathematics or physics (NM). The results are given in Table 2.

Table 2 Responses to the gravitational force task.

Response	LM (%) (n = 38)	UM (%) (n = 41)	UP (%) (n = 7)	NM (%) (n = 37)
↓ ↓ ↓ *	31	61	12	5
↑ NF ↓	45	24	38	43
↑ ↓ ↓	8	7	0	38
Other or no response	16	8	50	14

* indicates correct response NF means no force

Force is a fundamental concept in mechanics, and gravitational force is particularly important and basic. The percentages speak for themselves, but it is worth pointing out that the UM results are better than the LM results, and that the UP results are poor for a group of science students for whom mechanics is very important.

Such results concerning mechanics have been verified in many experiments around the world and are reported in Jagger (1985). It is clear that many of the fundamental concepts of mechanics are misunderstood or are subject to alternative conceptions. Perhaps the best documented of alternative conceptions is that force is proportional to velocity, and this incorrect view is very resistant to teaching. As with place-value, we have here a conceptual area which it has taken the human race thousands of years to master, throughout which period just such alternative conceptions as individuals hold today have been held by the experts of the day. It was not until Newton that the view that force and acceleration (strictly rate of change of momentum) are in proportion, not force and velocity.

Subsequently, later in a mechanics course, motion in a vertical circle is studied. The following question was used by Roper (1985).

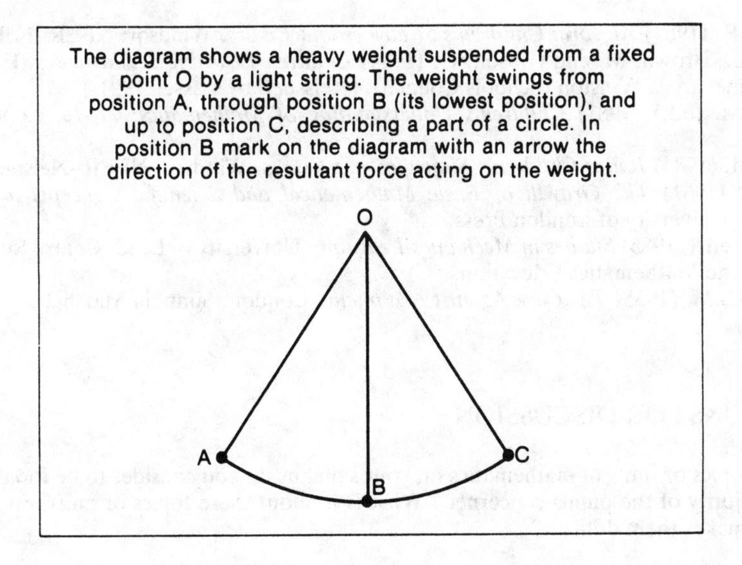

The results are given in Table 3.

Table 3 Responses to the force in circular motion task.

Response	LM (%) (n = 38)	UM (%) (n = 41)	UP (%) (n = 7)
↑ *	11	27	14
→	39	29	14
NF or W = T	13	17	0
↓	16	5	29
Other or no response	21	22	43

* indicates correct response
NF means no force

From these results we see that only a minority of students had mastered the implications of the concept of force in this situation of vertical circular motion. A considerable number of students believed that the force acts in the direction of motion, and this is another popular alternative conception which is resistant to teaching. Such alternative conceptions do not arise from direct teaching. They are brought to the classroom by the student. Our experiences of force and motion throughout life lead us to draw conclusions which may be incorrect. Intuitive beliefs extracted from our own experience explain mechanics for us in a way which we find acceptable and perhaps even helpful, but they may be wrong. Mechanics illustrates perhaps better than any other branch of mathematics the difficulty we, the teachers, have not just in trying to promote the learning of new ideas but also in trying to replace incorrect conceptual bases by correct ones.

SUGGESTIONS FOR FURTHER READING

Booth, L. R. (1984) *Algebra: Children's Strategies and Errors*. Windsor: NFER–Nelson.
Dickson, L., Brown, M. and Gibson, O. (1984) *Children Learning Mathematics*. Eastbourne: Holt, Rinehart & Winston (Schools Council). (Reissued by Cassell 1991.)
Hart, K. M. (ed.) (1981) *Children's Understanding of Mathematics: 11–16*. London: John Murray.
Hart, K. M. (1984) *Ratio: Children's Strategies and Errors*. Windsor: NFER–Nelson.
Lovell, K. (1961) *The Growth of Basic Mathematical and Scientific Concepts in Children*. London: University of London Press.
Orton, A. (ed.) (1985) *Studies in Mechanics Learning*. University of Leeds Centre for Studies in Science and Mathematics Education.
Renwick, E. M. (1935) *The Case Against Arithmetic*. London: Simpkin Marshall.

QUESTIONS FOR DISCUSSION

1 What topics or units of mathematics on your syllabus do you consider to be too difficult for the majority of the pupils concerned? What is it about these topics or units, do you think, which makes them difficult?

2 What aspects of work on fractions should be included in primary school mathematics and why?
3 How is a poor level of understanding of place-value (or ratio, or force) likely to affect pupils' ability to cope with your syllabus?
4 How does level of motivation affect the growth of mathematical understanding?

Chapter 3

What Cognitive Demands Are Made in Learning Mathematics?

THE PROBLEM OF CLASSIFICATION

There has been a variety of attempts to classify the mental activities involved in learning. Gagné (1970, 1977) listed and described eight types of learning. Bloom *et al.* (1956) analysed the objectives of education in the cognitive domain. Skemp (1971) discussed the processes which need to be adopted in doing mathematics. Polya (1945) attempted to analyse the process of solving mathematical problems, a theme subsequently taken up by Wickelgren (1974). Brown (1978) suggested that there were four types of mathematical learning, namely simple recall, algorithmic learning, conceptual learning and problem-solving. Her Majesty's Inspectorate (1985) listed five main categories of objectives for mathematics learning, and these were facts, skills, conceptual structures, general strategies and personal qualities. The four cognitive categories bear a close resemblance to those of Brown, and basically provide a suitable structure for further discussion, though in reality all four are inextricably linked in the learning process. In particular, 'the whole of cognition may be said to be a study of memory' (Claxton, 1984).

RETENTION AND RECALL

Children are expected to be able to recall from memory a variety of different qualities in mathematics, for example:

- words (e.g. length, metre, triangle)
- symbols (e.g. +, −, ×, ÷, /)
- numerical facts (e.g. number bonds, tables)
- formulas (e.g. $A = lb$, $A = \pi r^2$).

Memory has been the focus for considerable research effort by psychologists over many years. At one time it was believed that our powers of memory could be improved by exercising them, in other words by being made to learn *anything*—

relevant and useful or otherwise. Such an extreme view is not now acceptable, though the value of exercise might not be completely discredited. The modern view of memory is that it is a feature of overall intellectual capacity, and that different people might even have differing capacities as regards what kinds of knowledge or understanding can most readily be remembered. As with the processing powers of the brain, human capabilities in terms of memory have been studied from physiological perspectives. There is no doubt that the chemistry and physics of the brain might provide the ultimate answers to problems studied in educational psychology but we do not have many answers yet. Such physiological studies are therefore not considered to be within the scope of this book.

It should be pointed out that psychologists have expressed the view that we possess both short-term and long-term memory. More recently the concept of working memory has been introduced, and this might be important in mathematical learning. What we certainly wish to achieve is long-term storage together with ready recall. The problem is how to achieve this. Retention of knowledge has often been associated in the past with rote learning. Drill (repetitive practice) was thought to be the answer to the problem of fixing knowledge in the memory, though subsequent difficulties of recall suggest that drill does not often achieve its objective. The recent history of curriculum development in mathematics, however, reveals a clear, new view from innovators that the emphasis should be taken off memory work so, for example, formula books were provided for candidates in certain examinations. There is considerable doubt whether this movement carried along with it the majority of mathematics teachers. There is an obvious efficiency factor in having knowledge readily to hand—but there may be even more in favour of memorizing relevant mathematics. The view from psychology is that committing knowledge to memory is important in terms of efficient processing but at the same time rote learning without meaning is relatively unhelpful. Cockcroft (1982) included practice of skills and routines in the list of features of good mathematics teaching but there were also other features. Rehearsal is necessary but it is unlikely to be sufficient as 'the kind of learning machine that we are is one that thrives on meaning' (Claxton, 1984). In other words, retention and recall are easier if what is learned is meaningful in terms of the network of knowledge already held in the mind of the learner.

One difficulty in putting this view into practice immediately emerges. What do we do for learners at the very beginnings of mathematics when there is virtually no network of mathematical knowledge in the mind of the learner? How, for example, is the child to learn the symbols 0 to 9 and the corresponding words? There is clearly meaning to be learned in the ideas of 'oneness' and 'twoness' and so on, but the symbols and words are in a sense arbitrary and therefore have to be learned by rote. Even as a child progresses through mathematics some element of rote learning must remain, in particular in relation to certain words and symbols. Some words may be remembered more easily because they are used in everyday life, for example, 'length'. Other words like 'metre' and 'centimetre' are rarely used in everyday speech and need to be practised. Meaning is involved in the relationship between the lengths which the two words represent and the connection with the prefix 'centi'. The word 'triangle' would seem to be very meaningful in its bringing together of two ideas, but there must be considerable doubt as to whether this is helpful when the teacher first talks about triangles because the idea of angle is likely still to be relatively unformed. Symbols must frequently involve rote learning. Some require very careful discrimination, for example + and ×, and also − and ÷. In learning mathematics, and particularly in the early years, it seems inevitable that learning by rote or by simple association will be involved.

There are a number of other ways in which retention can be fostered. Simple devices such as variations in layout in text and exercise books, different type style, different colours, the placing of certain key elements in boxes and summary notes are all helpful. Repetition, or rehearsal, has a part to play, both spoken rehearsal and written rehearsal. Constant repetition of multiplication tables was once commonplace in mathematics lessons. Such learning techniques cannot be considered bad if they achieve their objective, but of course they often did not, and in any case there are relationships and properties within tables which give a conceptual component to them which suggests that repetition alone is not likely to be the only way of promoting learning of tables. Rehearsal, however, must not be rejected out of hand as a way of assisting in the fostering of retention of facts. Periodic revision, likewise, is also important.

Retention can also be promoted by using deliberate contrivances such as mnemonics. The use of a variety of mnemonics in learning the basic three trigonometrical ratios has been common, for example 'oranges have segments (sectors), apples have cores', for

$$\text{sine} = \frac{\text{opposite}}{\text{hypotenuse}}$$

and

$$\text{cosine} = \frac{\text{adjacent}}{\text{hypotenuse}}$$

or the more lengthy statement, 'some officers have curly auburn hair to offer attraction', which includes cues for the tangent ratio as well. It is interesting that mnemonics have not been used widely outside trigonometry. It may be that opportunities for using other mnemonics in the rest of mathematics are very limited. But they do work, and we must acknowledge that and use them as appropriate.

Even if retention is achieved we cannot test it without recall, and recall can be a serious problem. Sitting and thinking, hoping the elements will come back, 'racking one's brains', is frustrating and tiring. Often, however, the presentation to the learner of an appropriate cue 'jogs the memory', but it is how to arrange for the cue which is the difficulty. Memory is, to some extent, context specific, which is why our memory is sometimes 'jogged' by reconstructing the situation in which the original experience occurred. Teachers are willing to provide children with appropriate cues, but there comes a time when pupils might have to manage without external help. In the case of mnemonics, the rhyme provides the cue. In other cases a concept map fixed in the mind might help the child to follow the network to the required element, or might release a complete structure of elements once a few key ideas are remembered. Structure built into the retention greatly assists recall. Learning which has been achieved simply by rote and without a link into a network does not facilitate recall.

Recently, considerable attention has been given to the idea of concept maps. A concept map is simply a linked network of related elements of learning material. It can

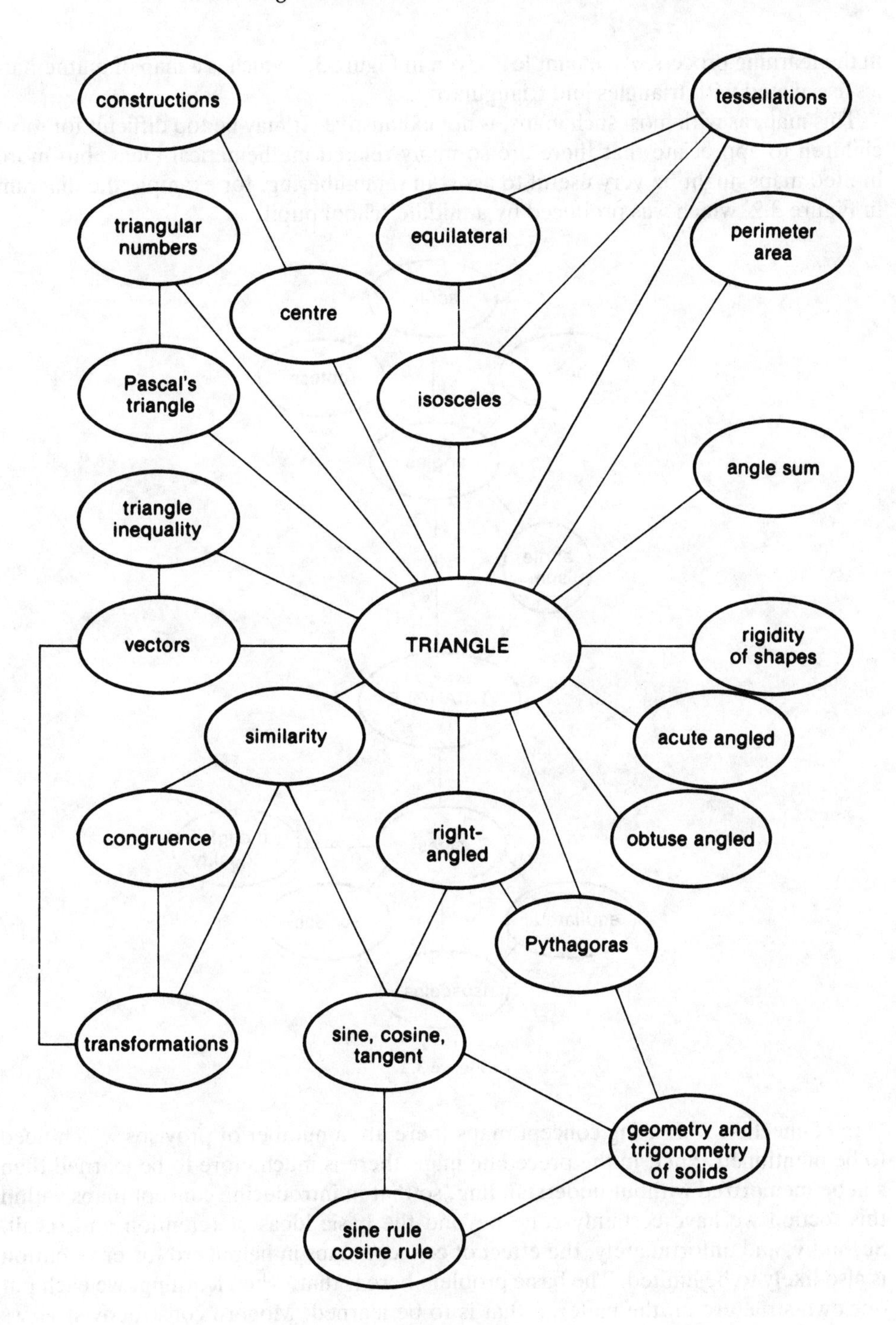

Figure 3.1

be used in a variety of ways. It can be used by teachers in course planning, it can be given to pupils as a model for revision, it can be used by a learner in a deliberate way in the learning process. An example is given in Figure 3.1 which is a map of mathematics associated with triangles and triangularity.

This map, as with most such maps, is not exhaustive. It may be too difficult for most children to appreciate that there are so many related mathematical ideas, but more limited maps might be very useful to assist in remembering, for example the diagram in Figure 3.2, which was produced by a middle school pupil.

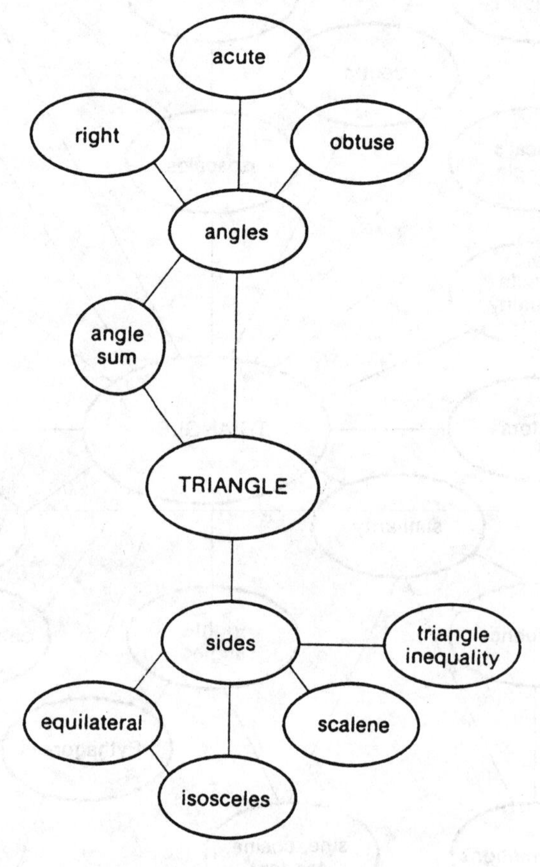

Figure 3.2

In connection with using concept maps there are a number of provisos which need to be mentioned. First, in the preceding maps there is much more to be learned than can be memorized without understanding, so that in introducing concept maps within this section we have certainly gone beyond the basic ideas of retention and recall. Secondly, and unfortunately, the effect of concept maps in helping to foster retention is also likely to be limited. The basic problem here is that, when learning, we each put our own structure on the material that is to be learned. Modern constructivist views suggest that learners do not remember material exactly as it was taught, that they construct their own meanings, that retention involves an active process of reconstruction. There is, therefore, the suggestion that many of the learning difficulties recorded in the previous chapter are not caused by failure to absorb all that was taught but rather that they are created in the reconstruction of knowledge. Retention and recall are clearly not simple processes.

USING ALGORITHMS

Learning mathematics is too much concerned with learning algorithms, it might be argued. The following are some examples:

- long multiplication
- long division
- adding and subtracting fractions
- multiplying fractions
- dividing fractions
- multiplying matrices.

Clearly, memory is involved in using algorithms, but here the children have to remember a step-by-step procedure. A worrying feature about algorithms in mathematics is that many which we expect children to remember and use with confidence are meaningless to the pupils, in terms of worthwhile knowledge, and are sometimes completely irrelevant. The distinction between instrumental understanding and relational understanding (Skemp, 1976) is helpful in appreciating this point, and this is illustrated below.

One of the less obvious algorithms to be found in some schools involves the conversion of denary numbers to binary. Assume that 13 is our denary number, then we divide it by 2 and record the quotient (6) and the remainder (1). Next we divide 6 by 2 and record the quotient (3) and the remainder (0). We continue until the quotient is 0.

```
2 ) 13 | 1
2 )  6 | 0
2 )  3 | 1
2 )  1 | 1
     0
```

Then the required binary number is 1101, formed from the remainders. It is possible for pupils to learn this as a procedure for conversion, but it is doubtful if many would appreciate why it works. Thus, they understand what to do to get the answer, so they have achieved instrumental understanding, but they have not necessarily achieved relational understanding. There is something of a parallel between this distinction and the distinction between memorizing by rote and memorizing through establishing connections.

There are many more well-known algorithms. Long division has traditionally, and probably wrongly, been taught in primary schools, and this was discussed in Chapter 2. Pupils in general would not be able to achieve relational understanding of long division and nowadays we would presumably use a calculator if we ever needed to obtain a quotient in a complicated division. The method for adding two fractions may be treated algorithmically:

$$\frac{a}{b}+\frac{c}{d}$$

$$= \frac{ad + bc}{bd}$$

However, this algorithm is sensible for $\frac{2}{3} + \frac{1}{4}$

but not sensible for $\frac{1}{2} + \frac{1}{4}$

or $\frac{1}{6} + \frac{2}{9}$.

Furthermore, there is a subtle difference in what one would sensibly do in the latter two examples which can only be appreciated through relational understanding, incorporating the idea of lowest common multiple. Recently, this algorithm has been linked with the idea of equivalent fractions. Since the equivalence of fractions depends on the equality of ratios it is open to question whether such use of equivalence has led to any greater understanding. In any case, why should we wish young children to be using a routine process to add fractions? The worst horror story of all concerning fractions involves division, for example

$$\frac{3}{5} \div \frac{7}{10}.$$

To find the answer, instrumentally, one 'inverts the second fraction and replaces the "÷" by "×"'. This has inevitably led to confused recollections, in particular which fraction to invert—or whether it is both. Relationally, of course, we wish to know how many $\frac{7}{10}$ there are in $\frac{3}{5}$, so equivalence is involved. Is the algorithm necessary? If it is necessary, at what age is it appropriate to aim for relational understanding? Such real understanding does not appear to be achievable by most pupils within the compulsory years of schooling.

A major problem with algorithms is that we often appear to introduce them before the pupils see a need for them. For example, we teach pupils how to solve linear equations by applying an algorithm to equations which can, and will, be solved by inspection, or trial and error. At the time of introduction to the procedure the equation

$$2x + 3 = 11$$

will not be willingly solved by the method

$$\therefore\ 2x + 3 - 3 = 11 - 3$$

$$\therefore\quad 2x = 8$$

$$\therefore\quad \frac{2x}{2} = \frac{8}{2}$$

$$\therefore\quad x = 4$$

when anyone can see almost at a glance that $x = 4$! Hart (1981) stated:

> We appear to teach algorithms too soon, illustrate their use with simple examples (which the child knows he can do another way) and assume once taught they are remembered.

> We have ample proof that they are not remembered or [are] sometimes remembered in a form that was never taught, e.g. to add two fractions, add the tops and add the bottoms.

One of the difficulties which faces us, however, is that we cannot be sure that relational understanding must precede the use of an algorithm. There is some evidence that relational understanding can be developed by thoughtful use of an algorithm, that instrumental understanding might help to promote relational understanding. Learning is so complex, and it seems like another chicken and egg situation. Which comes first? Nevertheless, there seems to be no doubt that too much instrumental learning is accepted in mathematics with pupils for whom relational understanding will never come, and that too much dependence on instrumental understanding in learning mathematics is rather like building a tower on insecure foundations. Such a tower will eventually crumble from somewhere near the bottom. As with so many aspects of learning, it is not easy to find the right compromise. This compromise might, after all, be different for different sorts of pupils! If we decide that a particular algorithm has some value is it justifiable to teach it, even knowing that relational understanding is impossible to achieve? Are there any essential algorithms anyway?

LEARNING CONCEPTS

There are problems in remembering facts in mathematics, and there are difficulties in learning algorithms meaningfully, but it is the conceptual structure or basis of mathematics which is perhaps the hardest aspect of all. Mathematics learning consists very largely of building understanding of new concepts onto previously understood concepts. Examples of concepts are so widespread that it is almost unnecessary to quote any, but for comparison with simple recall and algorithmic learning here are a few:

- triangularity
- percentage
- relation
- similarity
- limit.

Strangely, however, it is not easy to explain what a concept is. A dictionary might tell us that a concept is an 'abstract idea'. The definition by Novak (1977) is helpful, namely: 'Concepts describe some regularity or relationship within a group of facts and are designated by some sign or symbol.' However, it is probably easier to understand such a definition retrospectively, as it were, after thinking about particular concepts and what is involved in using them.

Skemp (1971) discussed exactly this point in a very helpful illustration of how we learn concepts. Considering the hypothetical situation of an adult born blind but given sight by an operation Skemp suggested that there is no way we can help the adult to understand the concept of 'redness' by means of definition. It is only by pointing to a variety of objects which are red that the adult could himself abstract the idea, the property which is common to all of the objects. Clearly, one would also assume that the counter-examples, the objects which were not red, would also help to clarify what was meant by 'redness'. Skemp was claiming that the learning of mathematical concepts is comparable. We must not expect children to learn through definitions. We need to use examples and counter-examples. Thus, in exactly the same way, we can run into difficulty in trying to define what we mean by a concept in mathematics unless we have many examples in mind.

The clear implication is that we learn about triangularity through examples of triangles and the contrast with other shapes. The concept of 'triangle' is probably relatively easy to grasp in this way, but we must not take it for granted. After all, children are sometimes very reluctant to admit that the shape in Figure 3.3 is a square,

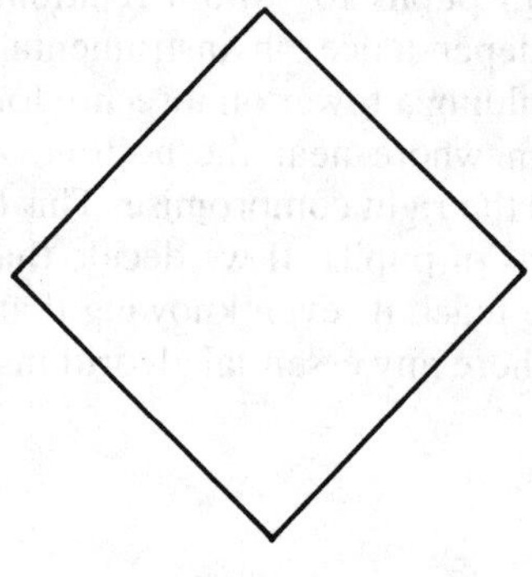

Figure 3.3

and often wish to call it a 'diamond' and insist that it is not a square. It would seem likely that our examples of squares, from which abstraction of the concept takes place, have not included a sufficient number for which one side is not parallel to the bottom of the page or blackboard. This is a point made by Dienes (1960) in connection with his theory of mathematics-learning. Other concepts, such as 'relation' and 'function', are certainly much more difficult to learn (see Orton, 1971). Usually, family relations or family trees are used as a way in, and they certainly help. However, a mathematical relation is, strictly, a set of ordered pairs. Relations in the family sense are the people (members or elements) to whom a particular member is related. The idea of a function in modern mathematics is very difficult to introduce, and probably irrelevant to most children anyway. Using functions without trying to define the idea abstractly may be the best starting point, and a more abstract definition of function can then be given much later only to those for whom it is appropriate. We need to be very careful when trying to introduce abstract mathematical ideas. Some ideas may be more abstract and therefore more difficult than we imagine. Skemp (1964), in drawing attention to the surprises we may get through concept analysis, cited fractions as being very much harder than we had previously thought and sets as being very much easier.

The precise mathematical definition of a concept, based on many years of handling examples, is something the mathematics teacher needs to have, but even that might cause problems. We all know exactly what a triangle is, but do we know what a natural number is? To many professional mathematicians the natural numbers are

0, 1, 2, 3, 4, 5 . . .

but to others they are

1, 2, 3, 4, 5 . . .

The definition of prime numbers at one time included the number one, and may still do so for some people, but nowadays most definitions of prime numbers exclude the number one. Some books might define rectangular numbers to include the square numbers, and others might not. Some books might define both 'rectangular' and 'oblong' numbers in order that rectangular numbers should include square numbers, with oblong numbers being those rectangular numbers which are not square, but other books might not admit to 'oblong' as being a useful idea. Yet it is possible to learn mathematics without having a completely tight definition of certain concepts. Certainly, our concepts grow and develop over the years. We introduce children to 'numbers', meaning only a very limited kind of number with which we count. Over the years other kinds of numbers are introduced, namely fractions (rational numbers), integers, irrational numbers and real numbers. Our original numbers have had to be redefined as natural numbers. Despite apparent complexities of concept redefinition which are implied, learning can still take place.

There are other problems too. Skemp (1964) expressed the view that sets were relatively simple to understand but how should we introduce the notion of *the* empty set? The emptiness of particular intersections of sets or of particular defined sets is an easy idea but not so the uniqueness of the empty set, nor the fact that the empty set is a subset of every set. Yet Skemp was correct. Learning about sets can take place successfully despite certain difficulties.

Boundary disputes might be less likely in mathematics than in other areas of knowledge, but they do exist, for example whether a square is a rectangle. Distinguishing between blue and green around the boundary between two colours might lead to disagreement, but is unlikely to hold up learning. We do not fully know what are the boundary problems for children in learning mathematics and how much they interfere with learning. The definition of concept by Child (1986) acknowledges the boundary problem:

> With most concepts there are wide margins of attribute acceptability . . . In some cases, the boundaries which distinguish concepts are hazy and ill-defined. But generally speaking, there is a large measure of agreement in the definition of most class concepts within a given culture.

The suggestion by Skemp that we do not learn concepts from definitions is not the only major strand of recommendation. In Nuffield (1967a) we find the simple proverb:

> I hear, and I forget;
> I see, and I remember;
> I do, and I understand.

Here is a strong activity message, also found in the Schools Council Report (1965) and in many other documents including Cockcroft (1982). In Cockcroft we find: 'For most children practical work provides the most effective means by which understanding of mathematics can develop.' The assumption in all of these references is that children, particularly young children, learn best by proceeding from the concrete to the abstract. Perhaps, to a large extent, we all learn in that way, though thinking purely in abstractions does become more possible, though not certain, in adulthood. Cockcroft, in many paragraphs, emphasized this kind of message, and suggested that it is important that we do *not* assume that practical approaches should be limited only to young children, or children 'whose attainment is low'. Cockcroft also stressed the slowness of the progression from concrete materials to abstract thinking. The Schools Council Report (1965) stressed the same, in: 'Children learn mathematical concepts more slowly than we realised. They learn by their own activities.'

Such views as those recorded above, and many other related views, are common to most publications which set out to make recommendations to teachers about how to help children to learn mathematics. The theories of learning which are discussed in some subsequent chapters of this book all face up to the crucial issue of how to promote concept learning. In fact, many of the above references show clear evidence of close association with such theories of learning anyway. It is interesting to note that one theorist, Gagné (1970, 1977), was at pains to suggest that some concepts can be defined. He suggested that there are 'concrete concepts' and 'defined concepts' and, whilst admitting that many concepts require a concrete approach, since they are fundamentally classes of objects, events and qualities (for example 'angle', 'triangle' and 'regularity'), he pointed to other concepts such as 'pivot', 'uncle' and 'sell' which cannot be learned from examples. This issue will be raised in the next chapter. Suffice to say at this stage that Gagné's view is open to debate and, if a distinction exists between concrete and defined concepts it is not a clear distinction, at least in terms of learning mathematics. What neither Gagné nor any other major theorist has denied is that, in the case of young children learning mathematics, attempts to define concepts are unlikely to be successful. Concrete approaches are often very necessary, though nothing will ever ensure that relational understanding is achieved.

Skemp and others have drawn attention to the implications for concept learning of what is perhaps best described as the hierarchical nature of mathematics. In some subject disciplines there might be considerable freedom as regards the order in which topics are taught. In mathematics it is usually much more important that we find the right sequence for the learner. Often the very examples which we use to promote concept learning are themselves other concepts, and we must be sure that these other concepts have already been adequately understood. A mature understanding of what we mean by 'number' as a generalization depends on an understanding of natural numbers, rational numbers, irrational numbers, integers and real numbers, together with, perhaps, an appreciation that this might not be a completely exhaustive list of different number sets. Various authors have tried to elaborate the hierarchy of concepts, or topics, through which learners must pass. Two such attempts are in Nuffield (1970) and Skemp (1971). It does not seem likely, however, that such hierarchies can ever completely solve our problems in the sequencing of learning in mathematics, though they should help. One other feature of learning is that we continue to refine and extend our understanding of concepts throughout life. A thorough and complete understanding of a concept is sometimes not only unnecessary to enable a learner to move on to the next concept, it might even be unattainable. It might be that it is the study of parallel or even more advanced concepts which leads to enhancement of the understanding of more elementary concepts. Another feature of learning mathematics is that some flexibility within the hierarchy of topics is possible, even though we have to be more careful with our sequence than in some other knowledge areas. Learners are not identical in their needs, after all, and do not all achieve identical levels of understanding of particular topics in a hierarchy. Hence the good sense of the well-known statement by Ausubel (1968):

> If I had to reduce all of educational psychology to just one principle, I would say this: The most important single factor influencing learning is what the learner already knows. Ascertain this and teach him accordingly.

PROBLEM-SOLVING

Considerable attention has been accorded in recent years to the place of problem-solving in mathematics and to how to help children to become better problem-solvers. It is first necessary to declare exactly what is meant by 'problem-solving' in this context. In fact, it is perhaps better first to say what we do *not* mean. At the end of a section of a mathematics textbook there is often a set of routine exercises, which may even be referred to in the text as problems, but these are unlikely to involve 'problem-solving' in the current accepted sense. The routine practice provided by such exercises is probably very important, and can be thought of in terms of rehearsal as a way of fostering retention in the memory. Some such exercises might require the learners to apply their mathematics to situations which arise in the real world and, as such, could be termed applications. Some such applications will involve problem-solving.

Problem-solving is now normally intended to imply a process by which the learner combines previously learned elements of knowledge, rules, techniques, skills and concepts to provide a solution to a novel situation. It is now generally accepted that mathematics is both product and process: both an organized body of knowledge and a creative activity in which the learner participates. It might, in fact, be claimed that the real purpose of learning rules, techniques and content generally is to enable the learner to do mathematics, indeed to solve problems, though Ausubel (1963) would disagree. Thus problem-solving can be considered to be the real essence of mathematics. Gagné (1970, 1977) has expressed the view that problem-solving is the highest form of learning. Having solved a problem, one has learned. One might only have learned to solve that problem, but it is more likely that one has learned to solve a variety of similar problems and perhaps even a variety of problems possessing some similar characteristics. Descartes expressed it as follows: 'Each problem that I solved became a rule which served afterwards to solve other problems.' It might therefore be asked, 'What is the difference between problem-solving and discovery?' Both require 'thinking', leading to the creation of something which the learner did not have before. Another term which is currently in frequent use is 'investigation'. An investigation might be closed, in the sense that the intention is to lead to an established mathematical result, or it might be open, in the sense that the result is not known in advance, or there might not even be a clear result which can be stated simply. Investigations should clearly lead to problem-solving. Investigations also, hopefully, lead to discovery. In short, whatever we mean by the separate terms 'discovery', 'investigation' and 'problem-solving', there are clear relationships between the processes involved. For this reason, a more detailed consideration of problem-solving is contained in Chapter 6, which is about discovery.

By definition problems are not routine, each one being to a greater or lesser degree a novelty to the learner. Successful solution of problems is dependent on the learner not only having the knowledge and skills required but also being able to tap into them and establish a network or structure. Sometimes a flash of insight seems to occur.

Although this is a phenomenon which is not fully understood, it usually involves the realization of some previously unacknowledged relationship within the knowledge structure. It therefore depends on having the richest possible knowledge base from which to draw. It is also known that it helps to turn the problem over in the mind thoroughly, to try out avenues of approach and thus to bring to the forefront a whole range of techniques and methods which might be appropriate. Further, it is known that the solution might still not come, but might come subsequently, after a period of time away from the problem, as if the subconscious mind, freed from the constraints of conscious attempts to solve the problem, continues to experiment with combinations of elements from the knowledge base.

There is considerable interest at the moment in aiming to improve the problem-solving skills of pupils in school. Polya (1945, 1962) has led the way in the consideration of how to establish a routine for problem-solving and, therefore, in how to train people to become better problem-solvers. Wickelgren (1974) too, basing his work on Polya but elaborating and extending this, claims to have evidence that his methods work and that they do produce more competent problem-solvers. It is interesting to speculate, however, that, although such training in problem-solving strategies might have considerable pay-off, we might be moving towards a more algorithmic approach, with all the inherent dangers of algorithmic learning. In fact, in contrast to Polya, Wickelgren and many current advocates of methodical approaches, Gagné has gone on record as stating that we probably cannot teach people to become better problem-solvers. This is because of his belief that one cannot teach thinking skills in a vacuum—each problem involves its own content and context, for if it does not, we have moved towards the routine exercises discussed at the beginning of this section. Having solved a problem we have learned something, but we have not become a better problem-solver *per se*. Ausubel (1963) too, whilst accepting that training in problem-solving within a fairly narrow and well-defined subject discipline might achieve some success, is at great pains to point out the transfer problem, raised in Chapter 1.

One aspect of problem-solving in mathematics is that often the problems are divorced both from the mainstream subject matter and also from the real world. Such puzzles may contain great interest for some children, but others may not see the point and be demotivated. Such puzzles are unlikely to produce knowledge or rules which are useful or applicable elsewhere. It has been a common feature of research into problem-solving and discovery that subjects have been presented with problems which are almost frivolous or whimsical. This has advantages in a controlled experiment, for it is most likely that all subjects start with the same knowledge of the situation—hopefully nil. It has also produced interesting results, but again we have the difficulty that whatever is deduced about problem-solving might not be transferable to more orthodox spheres of human knowledge.

SUGGESTIONS FOR FURTHER READING

Byers, V. and Erlwanger, S. (1985) Memory in mathematical understanding. *Educational Studies in Mathematics* **16**, 259–81.

Howard, R. W. (1987) *Concepts and Schemata*. London: Cassell.

Lindsay, P. H. and Norman, D. A. (1977) *Human Information Processing* (Chapters 8–10). New York: Academic Press.
Skemp, R. R. (1971) *The Psychology of Learning Mathematics*. Harmondsworth: Penguin Books.
Skemp, R. R. (1976) Relational understanding and instrumental understanding. *Mathematics Teaching* **77**, 20–6.

QUESTIONS FOR DISCUSSION

1. To what extent is the provision of formula sheets and books and other means of reducing memory load justifiable in mathematics?
2. How essential is relational understanding in learning mathematics?
3. Justify the algorithmic content of your mathematics curriculum.
4. Choose a unit of mathematics and analyse it in terms of memory load, algorithmic content and conceptual demand.

Chapter 4

Could We Enhance Learning Through Optimum Sequencing?

BEHAVIOURISM

How do children best learn multiplication tables? Do they learn best by chanting? Do they learn best by investigating number patterns from knowledge of addition bonds? Do they learn best by practising correct responses for given stimuli, such as random products presented on flash-cards? Or do they learn best through a mixture of methods? Different mathematics teachers are likely to hold different views on how children best learn multiplication tables. Chanting is still used, but as the predominant method of learning tables it has not been regarded with favour for many years. At the opposite extreme, investigation of number patterns and relationships may not fix products and factors in the memory. We want children to 'understand' why $7 \times 9 = 63$, but we also hope that the stimulus (perhaps presented on a flash-card)

$$7 \times 9,$$

will produce the instant response

$$63.$$

However one defines behaviourism, it is likely that some of the practices of teaching associated with behaviourist learning theories will be used in the teaching of elementary arithmetic such as multiplication tables.

How should we define behaviourism? Different authors appear to define behaviourism in different ways, so it is not easy to present a definition with which all interested parties would agree. Early behaviourist psychologists trained animals to exhibit required patterns of behaviour to prove that conditioning worked. One very well-known experiment was carried out by Pavlov, who conditioned dogs to salivate in readiness for eating on merely hearing the ringing of a bell. More recently Skinner conditioned rats and pigeons to perform required actions, usually in order to obtain food. Skinner (1954) said: 'Once we have arranged the particular type of consequence called a reinforcement our techniques permit us to shape up the behaviour of an organism almost at will.' It seems that Skinner was suggesting that what could be achieved with animals could be achieved with humans; people could also be conditioned to exhibit the requisite behaviour.

One problem with defining behaviourism is that it has not stood still. In a developing discipline one would expect that newer theories will extend or amend older theories. For this reason, reading the literature brings to light a number of technical terms, not only 'behaviourism' but also 'associationism' and 'connectionism', all of which appear to describe behaviourist-type beliefs. There are undoubtedly differences between the strict definitions associated with these technical terms, but it is not appropriate or necessary to discuss these here. A useful, simplistic definition of behaviourism as it is practised today is that it is the belief that learning takes place through stimulus-response connections, that all human behaviour can be analysed into stimulus and response.

It is also difficult to trace when the theory of behaviourism emerged just as it is difficult to say when calculus was invented. Thorndike was certainly an important, and early, instigator of ideas but perhaps one should look much further back, for example to Herbart one hundred years before Thorndike, to see where some ideas originated. A comprehensive consideration of the development of such learning theories will be found in Bigge (1976). Modern neobehaviourists, like Gagné, appear to hold very different views from those held earlier this century. Gagné (1977) expressed his views about learning as follows:

> Learning is a change in human disposition or capability, which persists over a period of time, and which is not simply ascribable to processes of growth. The kind of change called learning exhibits itself as a change in behaviour . . .

Whatever the best definition of behaviourism, an important belief running throughout its development has been in the effectiveness of stimulus-response learning. As a result of a particular stimulus the required response is elicited.

S leads to R

⟶

Given an appropriate question from the teacher, or from a book or programme, the correct answer is obtained. Learning proceeds, slowly but surely, through a sequence or chain of stimulus-response links. The effectiveness of the programme depends on the quality of the sequencing. Feedback, reinforcement and reward are often considered to have important places in the application of the theory. It might be a sufficient reward for a learner to receive instant feedback as to whether a particular response is correct, and this should then promote the desire to be presented with the next stimulus. A cycle of learning is thus generated, as depicted in Figure 4.1.

One of the early behaviourists, the American psychologist, Thorndike (1922), postulated a number of laws which have promoted discussion and debate ever since, and two of them are re-stated below in a concise form. Although these laws were proposed many years ago it is interesting to consider whether they are still acceptable today in the teaching of mathematics.

(1) *The law of exercise*
The response to a situation becomes associated with that situation, and the more it is used in a given situation the more strongly it becomes associated with it. On the other hand, disuse of the response weakens the association.

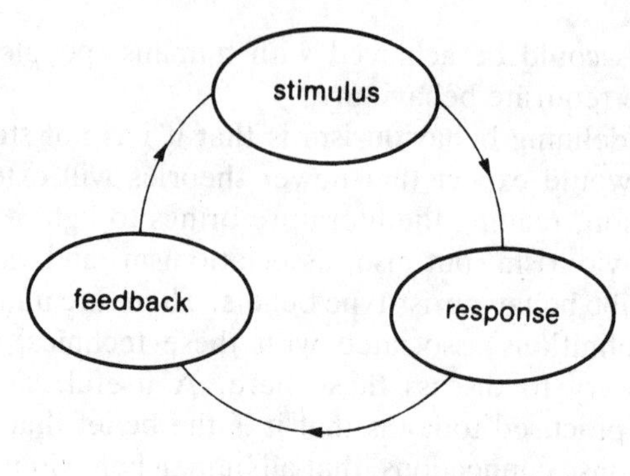

Figure 4.1

The retention of multiplication tables might be considered to fit this exactly. Further, much of the teaching of mathematics has consisted of the teacher demonstrating a method, process, routine or algorithm to be used in particular circumstances, followed by the class carrying out an exercise consisting of routine questions which can all be solved using the given process. Often the teacher will have incorporated question and answer techniques, but, nevertheless, the pupils will have been firmly steered in the direction the teacher wished them to go. The Cockcroft Report (1982) drew attention to this, and paragraph 243 has been much quoted. Exposition by the teacher followed by practice of skills and techniques is what most people remember when they think of how they learned mathematics. Even if we do not believe that 'practice makes perfect' we do believe that practice is the best way to strive after that perfection. We are seeking to establish a strong bond between the stimulus (the question-type) and the response (the application of the method of solution leading to the correct answer). This seems to be a direct application of the law of exercise. Teachers will know only too well, however, that most pupils subsequently appear to suffer severe weakening of this bond, assuming a bond existed in the first place. It is possible to teach pupils how to add two fractions together by practising the algorithm only to find that in the end-of-year examination, several months later, most pupils demonstrate only that they cannot respond correctly. The addition of fractions is a carefully chosen example, because we appear to re-teach it to many pupils every year from the moment they are first introduced to it around the age of 10. After revision, many pupils can once again cope; by next time round a year later many will have forgotten again. Disuse of the response might well appear to weaken the association. This apparent weakening of association would, however, be explained by cognitive psychologists in a different way from the behaviourists.

(2) *The law of effect*

> Responses that are accompanied or closely followed by satisfaction are more likely to happen again when the situation recurs, while responses accompanied or closely followed by discomfort will be less likely to recur.

Certain of the things we do as teachers also appear to suggest some acceptance of the law of effect. There are many ways in which satisfaction for a pupil can result from a response. Ideally, if a response is correct and the pupil knows it, satisfaction is achieved and the pupil is reinforced. However, many teachers seem to assume that it is necessary to provide satisfaction in an extrinsic way. Rewards are offered for good work; prizes, merit marks, gold stars and the like are common in education, on the assumption that they will perhaps maintain a pupil's interest and motivation or prompt other pupils to attempt to achieve the same. Poor work may result in punishment, certainly in low marks, producing discomfort and theoretically resulting in no recurrence of such poor work. The effects of such an approach to education are not all beneficial. Pupils often look for the line of red ticks or the complimentary comment rather than for written comments from the teacher intended to improve understanding and attainment. A line of red crosses is certainly a negative reinforcer; most children would wish to avoid this happening, but without help from the teacher there is little chance that the pupil will improve.

The reason for the title of this chapter is that it is a question with which the behaviourist approach to learning seems to have been closely involved. Two important examples of the careful sequencing of learning material are considered later in this chapter, namely programmed learning and learning hierarchies. However, before that, it is appropriate to look at the place of objectives in teaching and learning mathematics.

OBJECTIVES

It is common practice nowadays to plan lessons, topics and courses from a starting point of clearly stated objectives. This is quite a recent development in the sense that, although teachers have always had to have in mind their intentions for a lesson, these were often fairly vague, and may perhaps have been better described as aims. An appropriate current aim might well be 'To use Pythagoras' theorem', but this will need to be made more specific when put into practice in a particular lesson. Hence, an appropriate objective might be: 'Given the lengths of the two shorter sides of a right-angled triangle, the pupils will be able to calculate the length of the longest side.' The need for objectives in planning instruction was summed up by Mager (1975) in the words, 'If you're not sure where you're going, you're liable to end up someplace else—and not even know it'.

It is open to debate as to whether objectives are necessarily associated with the behaviourist approach to instruction. Gagné (1975) appeared to assume that objectives were part of behaviourism as he interpreted it, and explained the place of objectives as follows:

> To define and state an objective for learning is to express one of the categories (or sub-categories) of learning outcomes in terms of human performance and to specify the situation in which it is to be observed.

Mager (1975), however, said:

> During the early sixties we talked about behavior, rather than about performance. This . . . [was] . . . unfortunate. People were put off, thinking objectives necessarily had to have something to do with behaviorism . . . Not so. Objectives [do] . . . describe . . . behavior, [but only] because behavior is what we can be specific about.

Whether objectives are part of behaviourism is largely immaterial in the context of this book, but objectives do appear to form a relevant part of the efficient sequencing of material to be learned.

One way in which objectives have been associated with behaviourism is the fact that they were widely discussed in connection with the programming of learning, itself a product of the behaviourist approach to learning. Mager's own early book (1962) was entitled *Preparing Objectives for Programmed Instruction*. Lysaught and Williams (1963) included an extensive discussion of the preparation of objectives on which to base the programming of educational subject matter. Early work, based largely on logical reasoning, led to the drawing up of taxonomies of educational objectives, for example by Bloom *et al.* (1956) and by Krathwohl *et al.* (1964). Considerable attention has subsequently been given to methodologies for the preparation of objectives suitable for use by teachers.

Writers seem to be generally agreed on why objectives are important; they:

(a) provide the teacher with guidelines for developing instructional materials and teaching method;
(b) enable the teacher to design means of assessing whether what was intended has been accomplished; and
(c) give direction to the learners and assist them to make better efforts to attain their goal.

It is not surprising that, because of (a), considerable emphasis is often placed on objectives in the training of new teachers. It is also not surprising that, because of (b), it is now expected that new examination schemes will be prefaced by clear objectives for the course; not an easy thing to achieve. It is suggested in (c) that learners should be clearly aware of course objectives. It would be false to claim that this is always the case, and perhaps this is one way in which we are failing to capitalize on opportunities presented by using objectives.

What should an objective look like? Is 'the pupils will understand probability' a suitable objective? It may be, for certain purposes, though perhaps some people would prefer such a vague statement to be called an aim. Gagné (1975) differentiated between objectives according to the receiver of the information. For the teacher, objectives provide a basis for instructional planning, for the conduct of teaching and for evaluating pupils' learning at the end, so the objectives need to be devised with these three points in mind. For the student, objectives might be expected to contribute to motivation and to provide feedback at the end. Headteachers, chief examiners for external examinations, parents—all will require different sorts of objectives. An alternative way of differentiating is to distinguish between the aims of a whole course and the objectives for small units of learning experience like single lessons.

The distinction between aims and objectives might well depend on the words used, and considerable discussion has taken place about, for example, verbs which are open to many interpretations, such as 'know', 'understand', 'appreciate', 'enjoy', 'believe' and 'grasp', as opposed to other verbs which are open to many fewer interpretations, such as 'identify', 'calculate', 'sort', 'construct', 'compare' and 'solve'. Under no circumstances could 'be able to understand mathematics' be acceptable as an aim or objective, but 'develop a liking for mathematics' could be a very worthy broad aim, though not a behavioural objective. Suitable objectives might include: 'given any two natural numbers, each less than or equal to 100, the pupils will be able to write down the sum.' In this we have great precision, the teachers and pupils alike know where they are going, and the examiner knows what sort of question to set to assess the pupils' progress. The more detailed example of the difference between broad aims and more specific objectives given below is concerned with mensuration.

Aims for topic on mensuration

1 To develop an understanding of the mensuration of certain basic and hence a wide variety of composite shapes.
2 To ensure that pupils have an adequate knowledge of the appropriate units involved and that they know and can use relevant formulas.

Objectives for topic on mensuration

1 Pupils will know that the *perimeter* is the distance around the boundary of a shape.
2 Pupils will know that, in the case of a circle, the perimeter is called the *circumference*.
3 Pupils will know that the *area* is the amount of surface contained within the perimeter.
4 Pupils will know the names of, abbreviations for, relative sizes of and relationships between the units of length mm, cm, m, km.
5 Pupils will be able to measure lengths using the units mm, cm, m.
6 Pupils will know that areas are measured in square units and will know the names, abbreviations for, relative sizes of and relationships between mm^2, cm^2, m^2, km^2.
7 Pupils will be able to calculate the perimeters of rectangles and triangles and also composite shapes based on these.
8 Pupils will be able to calculate the areas of rectangles, triangles and composite shapes, using $A = 1 \times b$ (rectangle) and $A = \frac{1}{2} \times b \times h$ (triangle).
9 Pupils will be able to calculate the circumferences of circles using the formula $C = \pi d = 2\pi r$.
10 Pupils will be able to calculate the areas of circles using $A = \pi r^2$, and hence will be able to calculate the areas of composite shapes involving circles.

Even then, are these objectives specific enough? For example, with what degree of accuracy do we wish pupils to be able to calculate the areas? From what number sets will lengths be drawn—will they always be whole numbers or are fractions allowed? What degree of complexity will be involved in the composite shapes? How do we introduce the various formulas and should we formulate objectives concerned with the derivation of, for example, πr^2? Have we missed out other essentials which might cause problems when we try to teach? For example, do we need objectives concerned with clarifying how we use the formula $A = \frac{1}{2} \times b \times h$? Readers may be able to raise other questions.

One problem which has emerged, then, is that the derivation of complete, detailed, unambiguous and absolutely specific objectives is itself an elusive objective to have. As Gagné and Briggs (1974) said, in connection with their systematic approach to forming objectives:

> When instructional objectives are defined in the manner described here, they reveal the fine-grained nature of the educational process. This in turn reflects the fine-grained nature of what is learned. As a consequence, the quantity of individual objectives appli-

cable to a course of instruction usually numbers in the hundreds. There may be scores of objectives for the single topic of a course, and several for each individual lesson.

This is certainly the case with the objectives for mensuration given above. Such objectives really only make sense when they are allocated, one or two at a time, to individual lessons. The following list of objectives taken from Holmes *et al.* (1980) forms a good basis for discussion. Although an age range is specified (11–12 years), statistics and probability are now taught in most age ranges, and so the list is a relevant one for most mathematics teachers to consider. Are they appropriate? Are they clear and unambiguous? Are they sufficiently fine-grained to be classed as objectives? Are they appropriate to individual lessons?

Objectives for pupils taking O-level statistics (current syllabuses).
Year 1 (Age 11–12)

Pupils should be able to:
carry out a simple census to find facts from a small, well-defined population,
draw a random sample from a small population,
sample from distributions such as those given by throwing dice,
generate random numbers and use random number tables,
obtain their own data by counting and measuring and use other sources of such data,
draw up their own frequency tables by tallying and read them,
draw and read bar charts for discrete data and for continuous data with equal class intervals,
read pictograms,
draw simple pie charts,
read time series,
find the mode, median, mean and range of a small set of discrete data,
assign probabilities in the equally likely case,
assign probabilities to the random selection of one item from a finite population,
use finite relative frequencies to estimate future probabilities,
find the probabilities of simple combinations of elementary events by addition,
draw simple inferences from bar charts and tables.

One of the most detailed examinations of how to prepare instructional objectives has been carried out by Mager (1975), and the book is a very helpful guide for teachers who wish to think carefully about objectives. Issues such as the refinement of objectives through the use of additional conditions, criteria of acceptable performance, sample test items and the avoidance of pitfalls are all considered in detail. For example, a simple objective might be:
'The pupils will be able to multiply together two 3-digit numbers.'
The following forms an additional condition:
'Without the use of a calculator.'
An objective which also involves a criterion of acceptable performance might be:
'With the aid of a calculator, the pupils will be able to divide 1-, 2- and 3-digit numbers by 1-, 2- and 3-digit numbers and express their answers to 3 significant figures with a 90 per cent success rate.'
Sample test items are obvious:
'The pupil will be able to solve linear equations in one unknown, e.g. solve for x in the following,

$$(a)\ 3 + 5x = 15,$$
$$(b)\ 3x - 2 = 8.\text{'}$$

Objectives are now included with the syllabuses for many external examinations. An interesting exercise for mathematics teachers is to look critically at the objectives associated with familiar external examinations and to assess their value in the teaching–learning–assessment sequence. Some might be found to have very little value indeed!

PROGRAMMED LEARNING

One of the ways in which behaviourist-type approaches to instruction have influenced teaching methods has been in commending the programming of learning. Some people take the view now that the programmed learning craze has come and gone, and one would certainly not find much obvious evidence of it in today's schools. There was, however, considerable interest in it around the 1960s, and it would be wrong to write it off as one of the bandwagons of the day because the advantages sought through the programming of learning still have to be considered. Programmed learning in our schools in the 1960s was based on books or booklets, or perhaps on cheap, hand-cranked, teaching machines. Ideally, programmed learning needs to be machine orientated, but although purpose-built teaching machines did exist they did not reach the schools. Nowadays we have computers, so teaching programmes may be presented on screen and pupils' responses can be via a keyboard. The remaining major problem is writing the computer programs to present the material.

It must be pointed out, before moving on, that computers are being used in schools in many ways, not only in the old-fashioned sense of programmed learning. Computer-assisted learning is itself a huge field of study, and there are many modes possible of which the programming of educational material is only one. The computer may be used, for example, as a magic blackboard, to provide simulation exercises, to allow exploration and discovery and to provide a database for investigation and deduction, to name just a few possibilities. In the context of this chapter, however, the computer will only be considered in its role in programmed learning.

The boxed example on page 46 is from a programmed learning textbook by Young (1966), which illustrates some of the problems inherent in the construction of such books.

To highlight the problems we may ask a number of questions. Where, ideally, should the answers be revealed to the pupils? How can cheating be prevented? How is the interest of the pupil to be maintained? How can the programme cater for different abilities? What is the role of the teacher? And, beyond such practical problems, how valid is the theory which supports the programming of material in a form like this?

Lysaught and Williams (1963) began their excellent book on programmed learning as follows:

> In the Blue Ridge mountain city of Roanoke, Virginia, pupils in the eighth grade of the local schools in a single term all completed a full year of algebra, normally reserved for the first year of high school in other parts of the country, and only one child in the entire Roanoke school system failed to perform satisfactorily on a standardized examination. At Hamilton College, Clinton, New York, nobody fails the logic course any more; moreover, the average of grades has risen markedly. At the Collegiate School, New York City, a private elementary and secondary institution for boys that long has maintained high

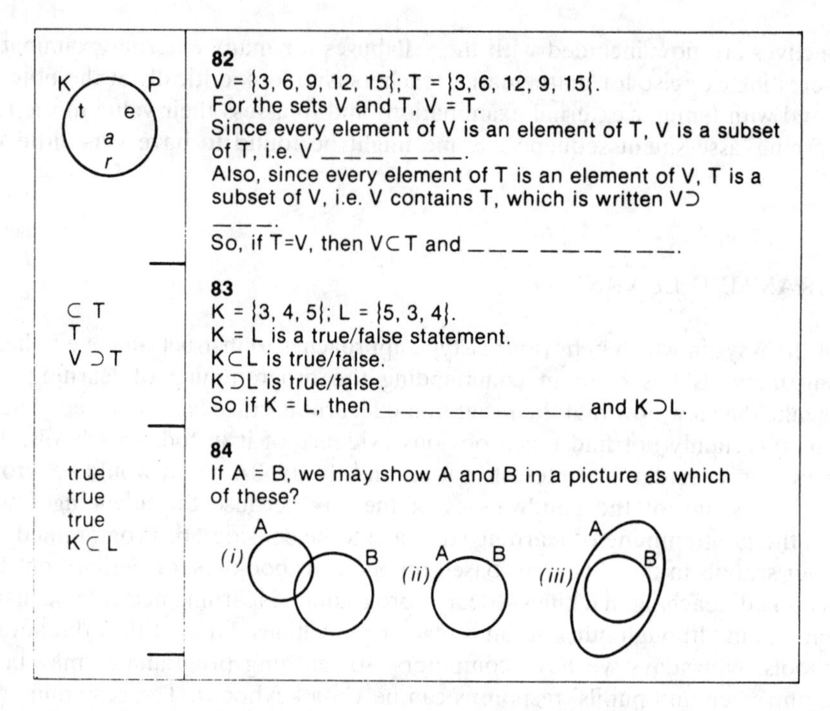

K P
t e
a
r

82
V = {3, 6, 9, 12, 15}; T = {3, 6, 12, 9, 15}.
For the sets V and T, V = T.
Since every element of V is an element of T, V is a subset of T, i.e. V __ __ __ .
Also, since every element of T is an element of V, T is a subset of V, i.e. V contains T, which is written V⊃ __ __ .
So, if T=V, then V⊂T and __ __ __ __ __ __ __ __ .

⊂ T
T
V ⊃T

83
K = {3, 4, 5}; L = {5, 3, 4}.
K = L is a true/false statement.
K⊂L is true/false.
K ⊃L is true/false.
So if K = L, then __ __ __ __ __ __ __ and K⊃L.

true
true
true
K⊂L

84
If A = B, we may show A and B in a picture as which of these?

(i) A B (ii) A B (iii) A B

Reproduced with permission (Young, 1966).

scholastic standards, students are now progressing more rapidly in modern mathematics than ever before. These diverse achievements are neither accidents nor as unrelated as a casual reading might suggest. Through them runs a common element. In each instance, classroom teachers have been utilizing the techniques of programmed learning, a method of pedagogy that increases the learning rate and proficiency of pupils and students.

There seems to have been considerable optimism, at that time, that the programming of educational material had an enormous amount to offer to school teachers.

The most famous name in the programmed learning movement in the USA, where it has been practised with perhaps more conviction than anywhere else, was that of the psychologist B. F. Skinner. One might say that programmed learning is based on conditioning, and Skinner's beliefs were confirmed for him through his enormous success in training animals through methods of conditioning. Many criticisms have been levelled at the theory for this reason. Why should we believe that because rats and pigeons can be conditioned to perform sometimes quite intricate movements in order to receive food that we can and should educate humans in a corresponding way? Yet Skinner believed that the possibilities for human learning were enormous. An important element of Skinner's theory was the practice of reinforcement, and the belief that, through reinforcement, the behaviour of even humans could be shaped at will.

Reinforcement has always been an important part of teaching methods. Early in the twentieth century reinforcement was largely based on fear: fear of incurring the wrath of the teacher, fear of punishment. Even today some aspects of the behaviour of children in school are based on their desire to avoid punishment or ridicule, and not on any desire to learn. Skinner was concerned that children were not learning in any positive sense, but that they were learning only to avoid the consequences of not learning. Dienes (1960) has also pointed out that learning in order to gain gold stars, house points, merit marks or even to gain a high position in the class is not good education. Praise and encouragement from the teacher can be a very good form of reinforcement but such praise is inevitably spread very thinly across the class of pupils so that any one pupil is unlikely to receive such reinforcement more than once in a lesson. Given the realities of life in today's schools it may be unrealistic to expect that all aversive forms of reinforcement and all extrinsic rewards will eventually be phased out. But the ideal of Skinner that all individual pupils will receive constant and rapid feedback of results and will, as a result, need no further form of reinforcement is not one that we can dismiss.

Teachers would be likely to agree that reinforcement, say through feedback of results to pupils, is important. Skinner's view would be that even this might not be enough if feedback were delayed, that even a matter of seconds between response and reinforcement could destroy all positive effects. Normally, in our schools, it takes many hours, sometimes days, before pupils receive feedback. This view of the importance of instant feedback is not universally believed today, and a contrary view may be that the quality and nature of feedback are what matters (see Hartley, 1980, for an introduction to this) but the belief in the immediacy of reinforcement was a part of Skinner's justification for programmed learning.

Skinner was also critical of the unskilled way in which pupils were introduced to new knowledge, and, in particular, the way in which they were expected to cope with sequences of material that were aimed at presenting groups of pupils with chunks of material. Such a criticism was, and probably still is, a very valid one. Each pupil ideally needs to proceed through a programme of work which is individually tailored to meet his or her needs. The theory is that each pupil requires that every step forward is small enough for that individual to accept. A possible counter-theory is that some pupils, at some times, may learn best through being plunged into a problem situation which is some way removed from their current state of knowledge and understanding; learning then takes place through finding ways of relating the new situation to the current state. Many teachers would feel that Skinner's theory is the safer one to accept, at least for many of the children they teach. It is, of course, impossible to say which is correct, as each might apply in different circumstances, or with different pupils; but these represent views which teachers have to acknowledge. Skinner's view influenced his approach to programmed learning.

It is interesting to record, in passing, that one of the conclusions that Skinner drew from his critical consideration of school learning had little to do with the programming of learning and was more akin to cognitive approaches to learning. Skinner noted that children gain reinforcement through practical approaches to learning, through interaction with the environment, through manipulating real objects. Some teachers would feel that rigid programmed learning is at the opposite end of a spectrum of learning styles from the active approach. Yet Biggs (1972) included programmed discovery as one of five different kinds of discovery in her commendation of discovery learning in the primary school. Perhaps programmed learning and active participation in learning are not irreconcilable.

In his explanation of programmed learning Skinner (1954) said: 'The whole process of becoming competent in any field must be divided into a very large number of very small steps, and reinforcement must be contingent upon the accomplishment of each step.' The boxed example is taken from Lysaught and Williams (1963).

S. *Quad* means four. *Lateral* refers to side. A *quadrilateral* always has four sides. A square would be one type of quadrilateral. The figures below are all ________ because they have four sides.

R. quadrilaterals

S. A rectangle [ ] is a quadrilateral because it always has ________ sides.

R. four

S. A rectangle is one type of ________ because it has ________ sides.

R. quadrilateral
four

S. A square is a type of ________ because it has ________ sides.

R. quadrilateral
four

S. All figures that have four sides are known as ________

R. quadrilaterals

Reproduced with permission (Lysaught and Williams, 1963).

In general, both appropriately small step-size and appropriate reinforcement are difficult to achieve for all pupils as individuals without the use of teaching machines. Skinner (1954) explained this as follows:

> As a mere reinforcing mechanism, the teacher is out-of-date. This would be true even if a single teacher devoted all her time to a single child, but her inadequacy is multiplied many-fold when she must serve as a reinforcing device to many children at once. If the teacher is to take advantage of recent advances in the study of learning, she must have the help of mechanical devices.

For the first time in the history of education such devices (electronic not mechanical) now exist in large numbers in the form of the microcomputer. Teaching machines did exist in the 1960s but were not particularly versatile or adaptable and, although used in the armed services, did not have any impact on our schools, presumably for reasons of cost. The hand-cranked 'machine' which schools had to be content with only survived the initial interest of the 1960s. However, the principles governing the presentation of material in programmed learning are the same whatever the means of presentation, whether by machine or by textbook.

Programmed learning involves the presentation of a sequence of stimuli to a pupil in the form of 'frames'. A single frame contains any necessary information and then poses a question which demands a response. The programming device used must provide a means for the pupil to make the response. A textbook, worksheet, hand-cranked 'machine' and some mechanical machines would very likely demand this response to be written on paper, on the page of a book or perhaps on a roll of paper in the machine. A modern computer is more likely to demand a response via the keyboard. An example of a single frame is given in Figure 4.2.

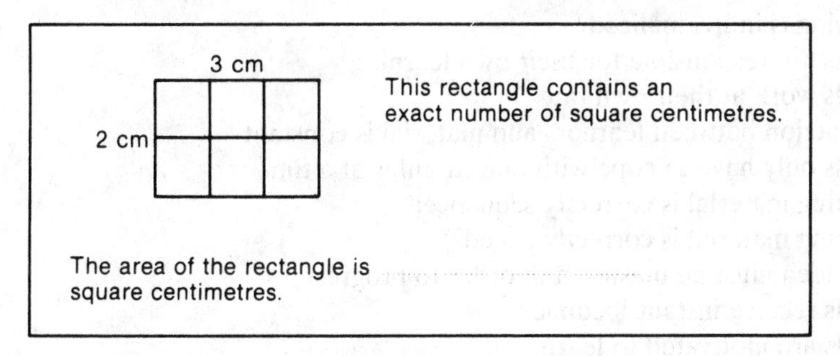

Figure 4.2

Having made a response, the pupil moves the programme along to the next frame, receiving feedback about the previous frame at the same time. The most elementary form of programmed sequence is a linear one, in which no deviation from a pre-determined sequence of frames is possible. A linear programme is clearly unsatisfactory in that there is little or no concession to the different needs of different pupils. With more versatile equipment variations on a linear theme can be introduced (see Lysaught and Williams, 1963). Using all of the above according to circumstances and to individual needs means a very complex, fully branching, programme is possible. Today's computers enable us to do this, but attempts to implement branching programmes in textbook form have not been all that successful.

The mode of computer-assisted learning based on programmed learning is often referred to as 'instructional'. Because free responses are difficult to cope with it is common to find that pupil interaction is via multiple-choice or 'yes/no' responses. In terms of branching, enormous flexibility may be achieved through using a computer. The computer acts as a teacher-substitute on a one-to-one basis, but pupils do not expose their learning problems to other pupils. An advantage for teachers is that pupils will not see a correct answer before making a response, which is something that cannot be guaranteed when using programmed learning texts.

Research into the value of instructional computer-assisted learning has raised questions about the assumptions behind the place of feedback. Simple knowledge of results may not be enough. What may be much more important is the location of errors and the provision of information to the pupil which allows such errors to be corrected. It may therefore be important, not that the step-size is so small that errors are avoided, but that the stimuli set out to bring misconceptions to light as a first step towards correcting them. Furthermore, feedback can be passive, merely informative, or it can be active in that it requires the participation of the learner. A common form of feedback is when the computer will not proceed until the correct key is pressed, but this has to be terminated sometime, and, if concluded by an instruction or message after which the next frame is presented, again the student can choose not to read it. Active feedback is in the form of further questions which demand response. The questions are intended to be those the student should be asking. One interesting study (Tait *et al.*, 1973) suggested that less able pupils particularly benefited from active feedback. For more able children it did not appear to matter.

To sum up, advantages claimed for programmed learning include the following:

- learning is individualized
- pupils are responsible for their own learning
- pupils work at their own rate
- interaction between learners and material is constant
- pupils only have to cope with one stimulus at a time
- learning material is correctly sequenced
- learning material is correctly paced
- each idea must be mastered in order to progress
- pupils receive instant feedback
- pupils are motivated to learn
- there is little problem of pupil anxiety
- a range of pupil abilities may be accommodated.

Such an impressive list should suggest to any experienced educator that there must be corresponding disadvantages. These would perhaps include:

- motivation generated by working with other pupils is missing
- inspiration generated by ideas from other pupils is missing
- pupil might choose to work too slowly
- pupil might unwittingly choose inappropriate routes through the programme
- material might not be sufficiently challenging
- material might hold some pupils back unnecessarily
- learning programmes are extremely time-consuming to prepare
- some kinds of learning experiences cannot be presented in programmed form
- there might be too much dependence on the honesty of the pupil
- material might lack essential interest and might not motivate
- learning might be promoted better when there is some anxiety in the pupil
- it might not be possible to accommodate the full range of pupil abilities.

As part of a curriculum, programmed learning has strong claims for a place. Some obvious uses are for individuals with special needs, for example enrichment for rapid workers, revision and repetition for slower workers and programmes to enable new pupils, or pupils who have missed work through illness, to catch up. Now that we have computers in our classrooms we need to carry out a continual appraisal of the place of instructional programmes in our teaching of mathematics. This, however, will not be the only way in which we will wish to make use of the particular advantages of the computer in the mathematics classroom.

LEARNING HIERARCHIES

Let us imagine that we wish to teach the multiplication of fractions to a class (why we should wish to do that is not our concern at the moment). Having defined our objective, for example 'the pupils will be able to find the product of any two rational numbers', we then have to determine what is our starting point for the sequence of instruction. What are we assuming that the pupils already know? In our list we might be tempted to include all aspects of work with fractions which traditionally come before multiplication. Our list might, therefore, look something like this:

1. products of natural numbers
2. knowledge that 'of' and '×' are equivalent
3. definition of a fraction
4. equivalence of fractions
5. lowest common denominator
6. how to change a mixed number into an improper fraction and vice versa
7. comparative size of two fractions
8. sums and differences of fractions.

Our list of prerequisite skills and knowledge, however, is intended to enable us to check off pupils' understanding and revise material where necessary. We do not want to be sidetracked into revising what is not essential in order for pupils to attain the objective, so the above list may contain irrelevant items. For example, do we really need to revise sums and differences of fractions in order to proceed to multiplication? Also, can we take it for granted that some items in the list are so elementary that they do not need to be revised? Products of natural numbers might fall into this category. Only when we have clarified both our objectives and our starting point can we begin to work out a sequence of instruction, which we might decide is as follows:

1. finding a proper fraction of a whole number
2. finding an improper fraction of a whole number
3. finding a proper fraction of a proper fraction
4. finding a proper fraction of an improper fraction
5. finding an improper fraction of a proper fraction
6. finding an improper fraction of an improper fraction.

But is this *the* correct teaching sequence? Are there alternative teaching sequences? Should 3 come before 2, for example? Which comes first, 4 or 5? How do we decide on the correct teaching sequence?

More than likely, teachers will make certain decisions as to the order of instruction according to intuition and preference. They may teach one order at one time and a rather different order at another time. But decisions of this sort do have to be made, and all of this comes before decisions concerning detail of how the instruction is to be carried out. Indeed, all of the steps along the road to the attainment of the ultimate objective may be regarded as prerequisite knowledge, each step requiring a clear statement of the objective. We could then hypothesize the existence of a sort of pyramid of stages, with the ultimate objective at the top, such as in Figure 4.3.

Without necessarily defining it in the above form, and without necessarily being

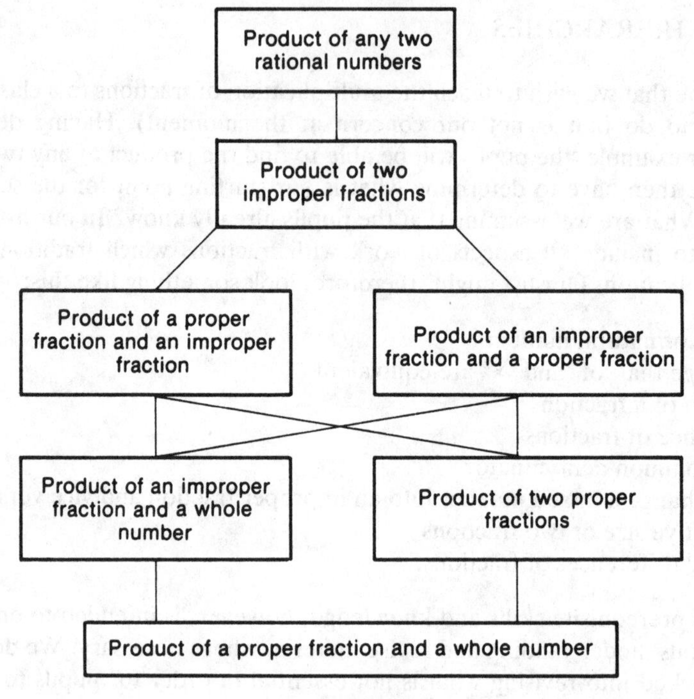

Figure 4.3

certain that the order is correct, a hierarchy of this kind is still very important for a teacher of mathematics to have in mind. If the teacher does not correctly identify the base line of the pyramid then some pupils will be lost right from the start. If the teacher does not identify all of the stages in the pyramid, and omits some, pupils will become confused somewhere in the middle of the hierarchy. If the teacher does not break down the steps into ones small enough for the pupils to cope with, many pupils will be unable to keep pace. Additionally, the teacher needs to check whether the objectives for each stage in the hierarchy have been attained before moving on to the next stage.

The theory of learning proposed by Robert M. Gagné is a more sophisticated and tightly controlled elaboration of this model. Gagné suggested that children learn an ordered, additive, sequence of capabilities, each new capability being more complex or more advanced than the prerequisite capabilities on which it is built. We have considered one such analysis, concerned with multiplying fractions. At a higher level one might hypothesize that in order to be able to solve quadratic equations by factors necessary prerequisites would include being able to solve linear equations, being able to find squares and square roots, and being able to factorize trinomials. At a lower level still there would be many more prerequisites concerned with ideas such as equality, products and quotients, sums and differences. A teacher moving on to solving quadratic equations might decide to revise some of the higher order or more advanced prerequisites but would probably take for granted that the children were able to cope with the more elementary prerequisites. These common-sense considerations suggest two other features of Gagné's theory, first, there is a variety of different kinds of prerequisite, some more advanced than others, secondly, the more elementary prerequisites can be ignored in devising the learning hierarchy.

A learning hierarchy, according to Gagné, is therefore built from the top down. We begin by defining the capability at the apex of the pyramid. This must be defined in terms of behavioural objectives, for example, 'pupils will be able to convert rational numbers in fractional form into decimals', or 'pupils will be able to find the sum of any pair of directed numbers'. The next stage is to carry out the detailed task analysis by considering what prerequisite capabilities are required in order to be able to attain the final capability (Figure 4.4(a)).

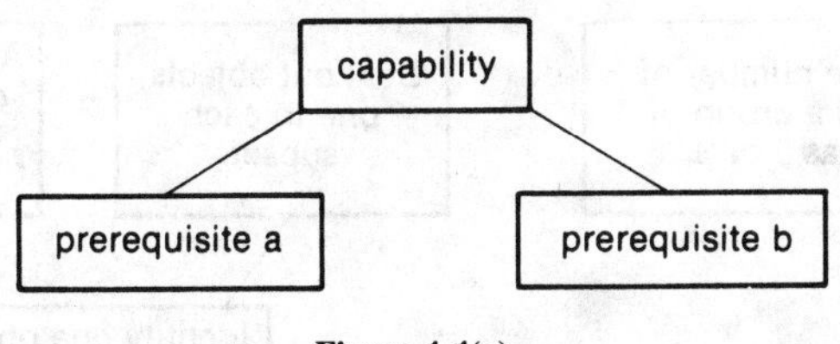

Figure 4.4(a)

We then repeat the procedure, by defining what prerequisites are required in order to attain prerequisites a and b (Figure 4.4(b)).

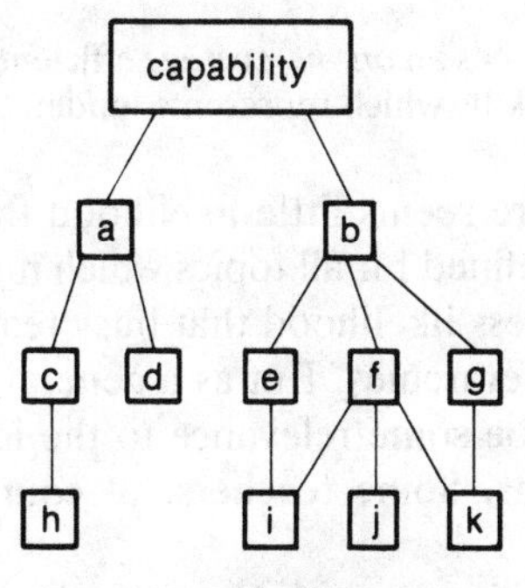

Figure 4.4(b)

Much of the research carried out by Gagné and his colleagues has been concerned with whether the hypothesized prerequisites were necessary and sufficient. If pupils possess prerequisites a and b can they always be taught the final capability? Can pupils who do not possess either a or b or both be taught the final capability? If pupils possess the final capability, is it always found that they possess both a and b? To carry this out at all levels of a hierarchy is very time-consuming, but Gagné's research has produced quite a number of such tested hierarchies, for example the one in Figure 4.5.

As one might expect in education, things do not always work out perfectly. For example, one is likely to find that there are pupils who do possess the final capability but do not possess either a or b or both. One also might find occasions when pupils can attain a or b without specific teaching in the process of receiving instruction on the final capability. One is therefore forced into the conclusion, as defined by Gagné (1977), that

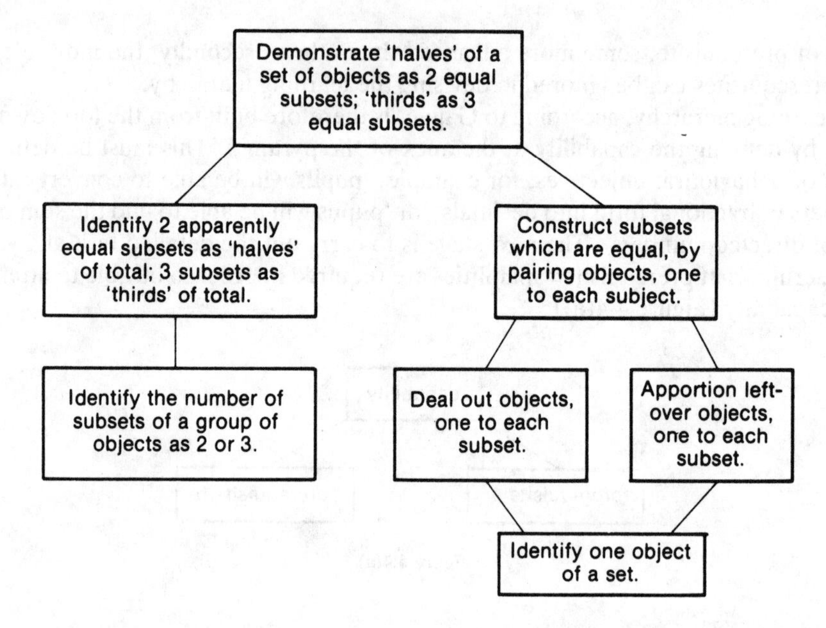

Figure 4.5 *A learning hierarchy for an early mathematical skill* (suggested by Resnick, 1967) (adapted from Gagné, 1970, reproduced with permission).

> a learning hierarchy . . . describes an *on-the-average* efficient route to the attainment of an organized set of intellectual skills which represents 'understanding' of a topic.

Another problem is that there seems little likelihood that tightly defined and tested learning hierarchies can be defined for all topics which might at some time be taught in mathematics. There is even less likelihood that busy teachers can involve themselves in devising research-tested hierarchies. But as a general idea, used in a more loosely defined manner, there must be some relevance to the idea of learning hierarchies in formal instructional situations. Some teachers, of course, do not often use formal instructional situations.

It is important, in a consideration of Gagné's views, to realize that his theory incorporates a view of 'readiness for learning' to which not all teachers would subscribe. Quite simply,

> developmental readiness for learning any new intellectual skill is conceived as the presence of certain relevant subordinate intellectual skills (Gagné, 1977).

Let us consider a typical conservation task (conservation tasks are described in more detail in the next chapter) with liquids poured from one shape of container into another, pictured in Figure 4.6. Gagné defined readiness according to the learning hierarchy in Figure 4.8 (see page 56).

A child is ready to learn any particular capability in the hierarchy, in this case conservation of liquid, if all the prerequisite capabilities have been mastered, and readiness depends on that alone. It is interesting to compare this view of readiness with that of Piaget. Piaget's view is that being able to perform such conservation tasks correctly depends on the stage of cognitive or intellectual development of the child,

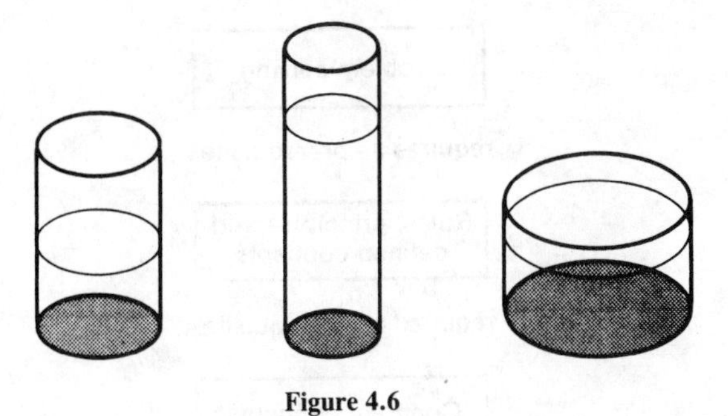

Figure 4.6

which is in itself defined by the fact that certain general logical processes have developed in the mind.

Gagné's theory of learning also incorporates views on the transfer of learning. The intellectual capabilities of a child do not remain specific; for example, if, based on appropriate learning hierarchies, learning of conservation has taken place with both rectangular containers and with cylindrical containers then the skills common to both situations will promote the generalizing of skills into other specific situations. Once a wide range of specific skills has been learned, for example in conservation, all other conservation situations will be that much easier to master. It then becomes possible to classify children as, in this case, 'conservers'.

The learning hierarchies of Gagné suggest that different prerequisites may be of different qualities, that there is, in fact, a hierarchy of types of learning. Let us consider Pythagoras' theorem, i.e. the sum of the squares of the lengths of the two shorter sides in a right-angled triangle is equal to the square of the hypotenuse (illustrated in Figure 4.7). The statement $a^2 = b^2 + c^2$ is clearly a rule of some kind

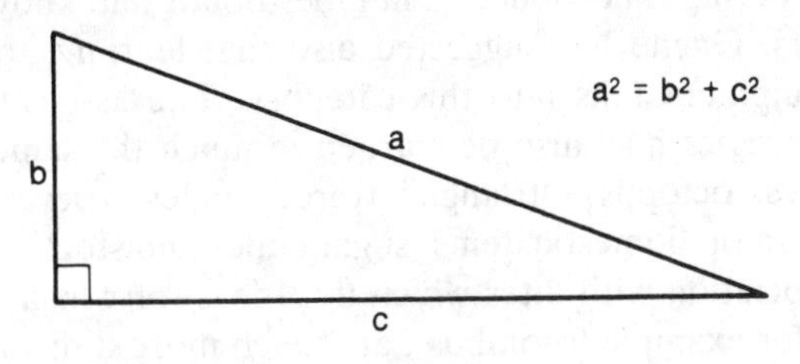

Figure 4.7

(which applies only to triangles when they are right-angled). A rule is a statement of a relationship between qualities. The relationship and the qualities both involve conceptual learning, for example squaring or area, equality, summation, triangle, right-angle, length, side, angle. The concepts themselves involve discrimination, between lengths and areas for example, and they also involve classification, what it *is* that is common to all triangles, for example. At quite a low level squaring involves products, and the most efficient way to find products is to know the multiplication tables. The learning of multiplication tables is likely to involve some elements of stimulus-response learning whatever one's beliefs about how tables should be learned.

It is therefore possible to draw up a linear hierarchy of types of learning which might apply to mathematics. The well-known list by Gagné is summarized in Figure 4.8.

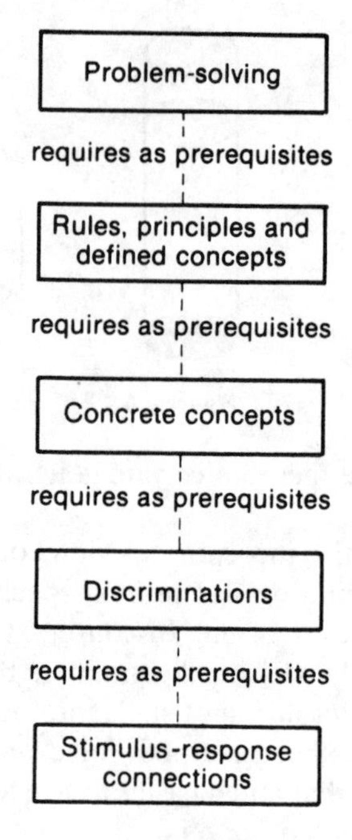

Figure 4.8

There are many examples one can give for each of these types as they relate to mathematics. At the lowest level much of the early number work might involve predominantly stimulus-response learning, for example learning number names and symbols and their ordering, knowledge of number bonds and knowledge of products, multiples and factors. Gagné has suggested also that learning to use mathematical instruments and equipment falls into this category. The association of names with ideas, objects or processes may also be learned in much the same way, for example 'octagon' (eight sides, octopus), 'triangle' (three angles, tricycle). 'Cube' may be learned from the common domestic items, sugar cubes and stock cubes. 'Kite' may be learned through association with kites which fly. Ideas for which ready-made associations do not exist, for example 'rhombus', are much more difficult to learn.

The ability to discriminate is as important in mathematics as in any other subject. From the very early stages of being able to discriminate between numbers of beads, counters or other objects held in the hand—is it five or is it six?—through to discriminating between Dx, dx and δx, mathematics is full of subtle differences and similar symbolisms. The symbols for the four basic rules, $+$, $-$, $\times$ and $\div$, possess remarkable similarities, so that if a teacher is careless in writing them the child can have great difficulty in discriminating. When children learn about different sorts of angles they have to be able to associate the correct term—acute, obtuse or reflex—with the correct angle. Many children struggle to come to terms with the difference between $2x$ and x^2, though there may be conceptual difficulties here as well as purely discriminatory problems. Properties of objects, for example the cube, need to be discriminated—edge, face, vertex, surface, area, volume, length, mass, weight. At a higher level still, such similar arrangements of letters and numbers as $3x^2$, Σx^2 and $5C_2$ need to be mastered and discriminated.

Mathematics is said to be very conceptually based, yet it is sometimes difficult to define exactly what we mean by a concept (Skemp, 1971). There are hierarchies of concepts too, concepts which are elaborated from other concepts being, perhaps, best described as of a higher order. One might claim, for example, that multiplication is a higher order concept than addition because multiplication, as repeated addition, depends on addition and cannot be learned before addition. In a different sense, number may be regarded as a higher order concept than natural number. It is necessary to learn what we mean by natural numbers, zero, fractions and decimals, negative numbers, integers, rational numbers, irrational numbers and real numbers in order to allow a mature concept of what we mean by 'number' to develop. Of course, young children will use the word 'number' when they mean, in the mathematician's terms, 'natural number', these being the only sort of numbers they have encountered. Their concept of number is a limited one.

There is probably an element of conceptual understanding in nearly all that we introduce in mathematics, from the beginnings of number work like multiplication tables and from elementary spatial work like identifying triangles and rectangles, through more complex ideas like sine and cosine to advanced procedures like differentiation and integration. The idea that new concepts can be learned fairly readily once prerequisite concepts have been mastered is not one which finds favour with many teachers of mathematics. The concept of place-value is very difficult for some children and yet comes so early in the logical hierarchy of mathematical concepts that it is difficult to accept that the problem only arises because prerequisite concepts have not been mastered. Learning to cope with place-value is a long, slow process for many pupils. At a higher level, the concept of ratio appears to depend on so little in the way of obvious prerequisites, and yet ratio and proportion are not mastered by many pupils by the time they leave school.

Principles are basically rules or laws. Gagné included 'defined concepts' within this category, that is, concepts which cannot be learned directly from concrete situations but require a definition. One example used by Gagné is 'diagonal'. He appeared to claim that it is necessary to state a definition, namely that a diagonal is a straight line which connects non-adjacent vertices of a polygon or polyhedron. Yet it is not always clear whether a concept is 'concrete' or 'defined'. Fractions (rational numbers) can be defined as ratios of integers; they can also be defined as parts of wholes (then mixed numbers are problematic); or they can be defined as operators (see Gattegno, 1960). Few people would claim to have learned what fractions were from a definition. The same is true with the concept of 'diagonal'—it is likely to be learned best from concrete considerations. The definition is a summary which comes later after many examples have been encountered. True principles may cause us less of a problem, particularly scientific principles like the gas law $P = r\frac{T}{V}$. Here we can clearly see that there is a relationship between concepts, though when children learn what the law really means it normally involves much more than being told about the relationship. It involves experimentation and measurement in order to observe that the relationship holds. It may be possible for more mature learners to accept principles without experimentation, but many teachers believe that children benefit from a concrete approach.

One might consider that one principle of mathematics is that 'equations remain valid if you do the same thing to both sides'. As a working rule for a particular stage of children's education this is very useful. It may present difficulties later (see Skemp, 1971), but many teachers use it around the 11–12-year-old stage. But again, it is doubtful that children would accept this rule as a statement without being able to construct it from concrete examples of what it means with familiar concepts like numbers. In terms of Gagné's analysis, it seems that we must classify Pythagoras' theorem as a principle or rule. But no teacher would introduce the rule without investigation, using numbers, squares of numbers and areas. The commutative law of multiplication is only defined as a rule once it has been found to hold in many numerical situations and one can generalize from it. So although principles (rules, defined concepts) might nominally be regarded as of a higher level of learning than concrete concepts this says nothing about the way principles might be learned.

Most people would agree that problem-solving might legitimately be regarded as the ultimate in terms of types of learning. It requires what we call 'thinking', and is dependent on a large store of knowledge and capabilities. One has first, however, to satisfy oneself that it is a form of learning. By problem we mean a question which requires some originality on the part of the learner for its solution, it requires the learner to put elements of prior learning together in a new way. Having solved such a problem, something has been learned (see Chapter 3).

Gagné's contribution to the study of how learning takes place and how it can be organized is a substantial one. In its entirety it may have few disciples in British schools, but it is worthy of study. Elements of it will turn out to be part of the approach of many teachers to lesson planning and presentation. In particular, the careful sequencing of material to be learned is likely to enhance the quality of learning. This sequencing, however, is not likely to be all that is required in the planning of learning experiences. The issue of what does need to be taken into account is continued in Chapter 5.

SUGGESTIONS FOR FURTHER READING

Gagné, R. M. (1985) *The Conditions of Learning* (4th edn). New York: Holt, Rinehart & Winston.

Hartley, J. R. (1980) *Using the Computer to Study and Assist the Learning of Mathematics.* University of Leeds Computer-Based Learning Unit.

Mager, R. F. (1975) *Preparing Instructional Objectives*. Belmont, CA: Fearon.

Skinner, B. F. (1961) Teaching machines. *Scientific American* **205**(5), 90–102.

QUESTIONS FOR DISCUSSION

1. What is the place of stimulus-response methods in learning mathematics?
2. Assess the learning objectives for a mathematics course which you teach in terms of comprehensiveness, specificity, lack of ambiguity and value.
3. Define the learning objectives for a unit of mathematics and prepare a draft learning hierarchy to guide teaching.
4. What is the value of instructional computer-assisted learning programs in mathematics teaching today?

Chapter 5

Must We Wait Until Pupils Are Ready?

ALTERNATIVE VIEWS

Once children have learned the meaning of addition and subtraction of natural numbers and are sufficiently skilled in carrying out the two operations, the thoughts of the teacher naturally turn to multiplication. Is there any reason why we should not press on immediately with multiplication? Are the pupils ready? Having learned about natural numbers and mastered all the standard operations on them is there any reason why we should not introduce our pupils to negative numbers and zero, and begin work on operations on integers? Is there any more to readiness for new mathematical ideas than adequate mastery of the mathematics which underlies the new ideas upon which they must be built?

The view that readiness for learning is simply 'the presence of certain relevant subordinate intellectual skills' (Gagné, 1977) was considered in Chapter 4 as being an interpretation of an aspect of behaviourist approaches to education. There are, however, alternative views around. Such alternative views have to be considered seriously when one acknowledges the learning difficulties experienced by pupils. If, for example, we treat fractions as an extension of the idea of number, as rational numbers in fact, it might be thought that pupils are ready for fractions once natural numbers have been adequately mastered. Yet many pupils struggle with operations on fractions for the whole of their school life, from the moment the ideas and techniques are introduced. Could such pupils really have been ready when we tried to teach them about operations on fractions? If we take an alternative view of the place of fractions in the curriculum, and regard them as part of a study of ratio and proportion, and acknowledge the difficulties inherent in a study of ratio and proportion, as illustrated in Chapter 2, it should not surprise us that doubts are raised about introducing operations on fractions as early as often we do. But what makes pupils ready for a study of ratio and proportion? Do many pupils struggle with ratio and proportion because we have simply failed to identify and teach all of the relevant subordinate intellectual skills?

The major alternatives to behaviourist views of readiness are developmental-type views. Simplistically, developmental approaches are likely to state that a pupil is only ready when the quality of thinking and processing skills available matches the demands of the subject matter. Furthermore, such thinking and processing skills are heavily dependent for their development on maturation, but also depend on environmental factors such as quality of schooling, home background, society and general cultural milieu. The required skills do not develop through maturation alone, though this is an important factor. It is more the case that interaction between the maturing child and all aspects of the environment make the required development possible. Developmental views achieved prominence through the publication of the work of Piaget and his colleagues at Geneva, and these views have become both better understood and modified with the passage of time. Many seeds of mathematics sown in our classrooms on the assumption that the earth is ready for cultivation appear instead to fall on stony ground. The developmental view needs to be considered and the best place to begin is with a study of what is relevant from the work of Piaget. It needs to be stressed at the outset, however, that although Piagetian theory may be interpreted as saying something about readiness, the readiness issue is not the most important issue discussed within the writings of Piaget.

PIAGET AND READINESS

Piaget's theory of intellectual development was based on results from experiments with children using the clinical or individual interviewing method described in Chapter 2. These experiments were carried out over a long period of time, commencing some fifty years ago, and were very numerous and varied in both nature and content. Many of the experiments were based on mathematical content and concepts and the results may be considered to throw light on the growth of understanding of mathematical ideas. Piaget himself came from a background in the biological sciences, and his view of learners as growing and changing organisms is the underlying basis for his developmental approach to learning theory. It should be pointed out that Piaget never claimed to be a learning theorist as such, but the extent to which his theory has been applied to education suggests that it is not inappropriate for others to view his work as providing theoretical bases for learning.

Piaget's experimental work was so extensive that it is possible only to select a very small part to illustrate both the emergence of the theory and the importance for learning mathematics. A suitable part is a selection of experiments concerning conservation.

One conservation experiment was based on containers of beads. Two equal amounts of identical beads were counted out into two identical containers, thus reaching the same level in both containers which were intended to be seen by children as equal in every sense. The beads from one of these two containers were then tipped into containers of a very different shape, first into a container which was both wide and shallow, and secondly into a container which was both tall and narrow (see Figure 5.1). At each stage of rest the child was asked whether there were the same number of beads in the two containers currently holding the beads. Piaget, and others who have repeated the experiment subsequently, noted that many children gave responses which, to an adult, would be considered strange, unexpected and certainly incorrect.

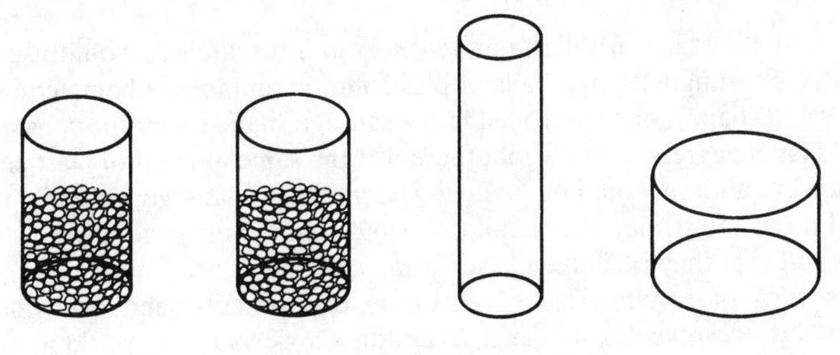

Figure 5.1

The younger children tested tended to express the view that the number of beads changed according to the shape of the container, that, for example, there were more beads in the tall, narrow container than in the original. What is more, when tipped back into the container in which they had first been placed there was once again the same number of beads as in the untouched original container. Testing children at a variety of ages led Piaget to conclude that children's first ideas are likely to be to insist that the number of beads changes, in fact to deny conservation. Eventually, after a period of time which can be confusing to the observer (in that the children sometimes accept conservation and at other times do not), the adult view is finally expressed and conservation is consistently admitted. As long as no beads drop out, there is the same number of beads no matter what shape the container.

A different experiment, but another approach to conservation, was to use liquids instead of beads, thus involving continuous pouring rather than discrete tipping. It has become common for researchers replicating this kind of experiment to use lemonade or orange squash in order to ensure capturing the full attention of the child and in order to motivate, perhaps through promises of a drink later. The questioning procedure, however, was always basically the same as for the beads. Again, identical quantities of liquid were poured into two identical containers. Alongside were one or two containers of different shapes, taller and more narrow than the original, or wider and more shallow (see Figure 5.2). Pouring the liquid into these containers of other

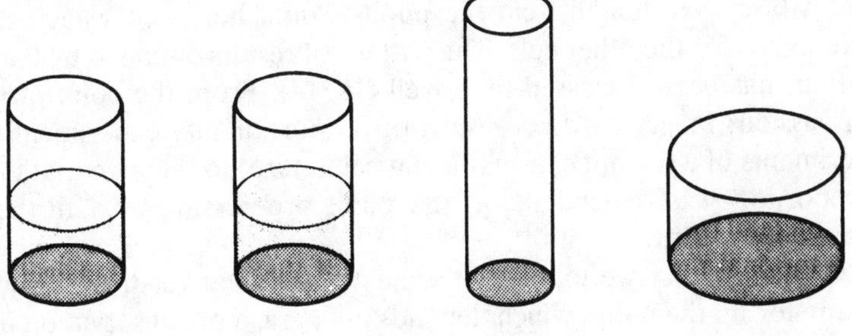

Figure 5.2

shapes was likely to produce the same responses as for the beads suggesting that the younger children were not accepting conservation. Older children, however, gave responses which were of the kind one would expect an adult to give.

A third version of the conservation experiment was originally based on the manipulation of modelling clay. Modern-day teachers and researchers would use plasticine for this task. Starting with two balls of plasticine, acceptable as being equal in every way (mass and shape), one was rolled into a sausage shape under the constant gaze of the child. Having agreed initially that there was the same amount of plasticine in both balls, younger children would be likely to claim that the sausage, or perhaps the ball, contained more plasticine. Older children would, however, generally give responses which suggest that they accepted conservation.

Piaget's view of results from conservation experiments, and from many other experiments which revealed a change in children's views of the world at around the same age, was that the structure and nature of intellectual behaviour had changed, that the children had moved on to a significantly different stage of intellectual development. In the case of conservation the critical age appeared to Piaget to be around seven years of age, though one naturally would expect some variation, so that for some children it might be around six years of age and for others around eight. The acceptance of conservation was not the only change in intellectual capability at around this age, suggesting that the critical feature was not some specific new understanding, like what happens when plasticine is moulded, but was, in fact, a wholesale change over a comparatively short period of time into a radically different stage in intellectual development.

Piaget suggested that there were a number of radically different stages in intellectual development. Thus he also used experiments which he claimed revealed a significant change in the nature and quality of thinking at other times in life. A number of experiments appropriate to a change later in childhood, in adolescence really, involved the ideas of ratio and proportion. One such task was adapted and used subsequently by Hart (1981) and was based on eels whose lengths were in a known ratio and whose appetites were proportional to their lengths. In one version of the experiment the eels were fed with 'discrete' items of food, meat balls for Piaget and sprats for Hart, and the pupil was required to calculate the number needed to feed each eel. Thus, given eels of lengths 5, 10 and 15 cm and given that the 10-cm eel needed four meat balls, the pupils would be asked how many meat balls each of the other eels would need. In another version of the experiment eels were fed with 'continuous' items, biscuits for Piaget and fish fingers for Hart. Given that the 10-cm eel needed a biscuit of length 6 cm the pupils would be asked what length biscuit would be required for the other eels. The pattern of results obtained by Piaget, and its interpretation, has been discussed by Lovell (1971a). From the point of view of the present discussion, Piaget used his results of proportionality experiments and many other experiments of a scientific or mathematical nature, to theorize that the ability to handle proportion was dependent on the pupil progressing to a further stage of intellectual development.

From Piagetian theory we may therefore extract a clear readiness view. Children are not ready for mathematics which depends on a grasp of conservation if they have not reached the stage of intellectual development at which conservation is accepted as a part of the way the world works. Pupils are likewise not ready for mathematics which is based on ratio and proportion (and there is a great deal of school mathematics in this category, including for example rational numbers and trigonometry), if they have not reached the stage at which proportionality has been mastered. Well-informed readers might feel that the above outline of aspects of Piagetian theory and the link with the readiness issue has been oversimplified. It is, for example, perhaps unlikely that Piaget would ever have presented the issue as it has been presented above, if only because the interpretation of his work in terms of implications for the mathematics curriculum has always been left to others. The justification for presenting the issue so simplistically is that it is precisely the way others have interpreted the results. Whether such a simplistic view is tenable now should emerge throughout the rest of the chapter.

It is now appropriate to mention briefly the stages of intellectual development which Piaget proposed. For convenience four major stages are listed below, though some of these stages incorporate substages, and some stages have subsequently been allocated simple subdivisions by other researchers and educationists following the same theoretical tradition. In fact, different reviewers of Piaget have grouped the stages and substages in slightly different ways, so it is possible to find authors referring to Piaget's five stages, or to four stages, or even to three. The four stages included here are:

1. the sensori-motor stage
2. the pre-operational stage
3. the concrete operational stage
4. the formal operational stage.

Piaget himself referred to a pre-conceptual stage and an intuitive stage, as subdivisions of the pre-operational stage. Such subdivisions as early concrete, late concrete, early formal and late formal have been used from time to time by others. All children, according to Piaget, pass through these stages and in that order, that is, they successively reveal those characteristics of intellectual activity which Piaget has spelled out for the stages.

From the point of view of learning mathematics the consequence should be that, if a child is known to be operating at a particular Piagetian level, if it is known at what stage they are functioning, there is no possibility that they will be able to cope with any mathematics which depends on capabilities associated with a subsequent stage. Acceptance of conservation is not a characteristic of the child's thinking before the concrete operational stage. Indeed, according to Piaget, a number of thinking skills emerge and develop with the onset of concrete operational thought, including class inclusion, reversibility, combination and separation, arranging in order and relative position, all of which might be very important in moving from an informal and intuitive approach to mathematics, involving little more than the manipulation of objects and materials, to mathematics as a paper and pencil activity. A major problem with this interpretation, however, is how do you identify at what stage a particular child is operating? Is it, indeed, possible to identify a stage and label a particular child in this way? It is interesting to note that the age of seven, as a suitable average, has been given particular significance and importance in English education. Without the benefit of Piagetian theory suggesting an important intellectual development at around the age of seven we traditionally classed children under seven as infants and children over seven as juniors. What is more, it was not unusual for the approach to learning to change radically with the change of school. Infants spent their time playing, but juniors got down to proper arithmetic with pencil and paper!

The two separate components of the term 'concrete operations' both require comment. The term 'operation' is common to three of the Piagetian stages listed earlier and, to Piaget, 'operation' possessed a precise meaning. Operations were to be thought of as actions, but carried out in the mind, and the operations were organized into a system. At the concrete operational stage these operations included combining, separating, ordering and so on, operations which have been described earlier. The term 'concrete' must not be thought of as implying that mathematics teaching always requires concrete apparatus until, of course, the full emergence of formal operations. The concreteness of the operations depends as much on actions carried out in the mind on the basis of prior knowledge of, and familiarity with, relevant underlying concrete manipulations. Thus, in a new learning situation, physical activity with actual objects is likely to be important at the concrete operational stage, but only up to the time when the child is able to replace such actual physical manipulations with corresponding mental activities. Concrete referents are always likely to be important at the concrete operational stage but should not be required all the time. On the whole, the evidence available suggests that concrete referents are not made generally available to pupils at around this age. Although there is sometimes misunderstanding about the relationship between the term 'concrete operational' and the use of concrete apparatus in the classroom, the usual error made in mathematics teaching has not been to overuse apparatus, it has been not to have apparatus as reference material sufficiently often. There are, of course, many people who cannot accept that such stages of intellectual development as described by Piaget, like the concrete operational stage, have any meaning. However, the need for concrete referents in teaching mathematics to most pupils for much of their school life does appear to exist independently of any belief or otherwise in Piagetian theory.

In terms of learning mathematics, the ability to cope with abstractions would depend on the emergence or development of formal operational thinking. Apart from proportionality, there are many mathematical topics and ideas with which teachers know that their pupils will have major difficulties because of the level of abstraction required. The whole of algebra as generalized arithmetic is dependent on abstraction from the rather more concrete numerical relationships. It is well known that algebra is found to be difficult, and indeed irrelevant, by many pupils, and some develop such an intense dislike of it because of this that it colours their whole attitude to mathematics. To these pupils there is no real meaning underlying what we ask them to do in algebra. Perhaps many pupils are not ready in the sense that we, the teachers, are always eager to press on to the next topic and we introduce algebraic ideas too soon and too quickly. The Piagetian explanation of this phenomenon would be along the lines that the development from concrete operational thinking to formal operational thinking is not sufficiently advanced at the time we wish to move ahead to the next algebraic ideas.

In terms of Piagetian theory it is only at the formal operational stage that one might expect dependence on concrete referents to recede into the background. We know that our most able pupils have little need for concrete apparatus as they move into and through the secondary school though, again, we are always likely to assume that they need equipment less than they perhaps do. Eventually the manipulation of symbols as an abstract exercise does become possible for a proportion of our pupils, but only for a small proportion. It seems that the majority of pupils are never ready for most of the algebra we would like to be able to teach. Formal operational thinking, to Piaget, allows hypothesis and deduction, it allows logical argument, it allows reasoning in verbal propositions. It is important to emphasize, however, that these more adult intellectual pursuits only become possible with the onset of the formal operational stage, they do not become certain. We all frequently need to function at a more concrete level, and often a practical introduction to a new idea is helpful, like using interlocking cubes to investigate number sequences and sums of series.

The implications for mathematics-learning are clear. Many mathematical ideas require the kind of thinking skills which Piaget has claimed are not available until the onset of the formal operational stage. It does not matter how carefully and systematically the teacher might try to build up a pupil's capabilities and knowledge—it is impossible to introduce concepts dependent on formal thinking before the pupil has reached that stage. The pupil is not yet ready for such abstract ideas. Pupils might, of course, be able to grasp the beginnings of an abstract idea in an intuitive or concrete way, but they cannot appreciate the idea as the teacher does. Explanations by the teacher will fail to make any impact unless such explanations are dependent only on skills available to pupils at the concrete operational stage.

Up to the present moment no attempt has been made to state the age at which children move from one Piagetian stage to another, apart from the references to 'around seven' for the arrival of concrete operational thinking. There are many problems with trying to relate stages with ages. Clearly, pupils do not pass suddenly from one stage to the next. There must be a period of transition between any two stages. One of the problems is that it might appear that children are in transition for much more of their childhood than they can be said to be identifiably operating within a particular stage. The idea of continual transition cannot be reconciled with Piagetian stage theory. Another problem is that it is difficult to identify categorically at which stage a particular pupil is at a particular moment in time. Yet it is interesting to speculate on how the Piagetian theory of intellectual growth compares with physical growth in terms of height. In this physical sense children do have periods of rapid growth and other periods of much slower growth. For example, before puberty many children appear to remain around the same height for a considerable period of time. In adolescence there is often a period of amazingly rapid growth before it slows again as adulthood is approached. It almost seems as if there are stages in physical growth. It must be a possibility that children experience periods of rapid intellectual development and periods of much slower development. This need not, however, imply that such relatively stable stages in intellectual growth are qualitatively different in the way that Piaget has suggested. Nor is there any evidence to suggest that changes from one Piagetian stage to the next coincide with periods of rapid physical growth.

Since Piaget first outlined his theory that intellectual development takes place in stages, and at the same time related the stages to ages, literature has been dogged by too optimistic a view of the rate of development in most pupils. Piaget suggested that the development from pre-operational thinking to concrete operational thinking took place around the age of seven, but there must be wide variation from pupil to pupil. Further experiment and research has suggested that those characteristics which Piaget described for the formal operational stage do not begin to emerge until 14 or 15 years of age in many pupils. This is much less optimistic than Piaget's original suggestion, which seems to accord only with evidence as regards very able pupils. There are undoubtedly differences between pupils of the same age and categorical statements about the age when pupils move from one stage to the next are not helpful. Cockcroft (1982) drew attention to this feature when he suggested a seven-year difference for a particular place-value skill (referred to in Chapter 2). The only description of intellectual development which would make sense to most teachers is one that takes this seven-year difference phenomenon into account. Figure 5.3 illustrates the likely re-

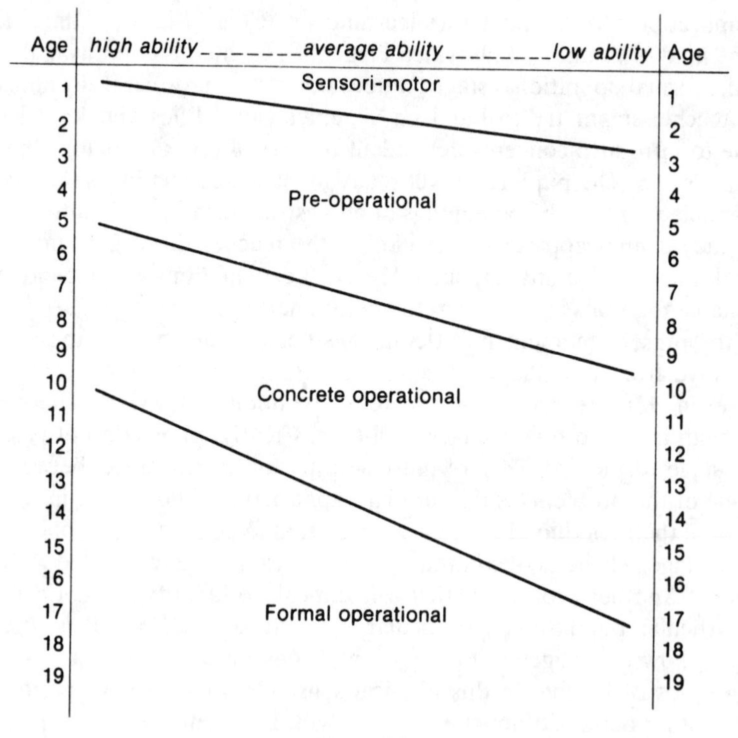

Figure 5.3

lationship between Piagetian stages and ages through the eyes of a mathematics teacher, though this diagram must not be taken as prescriptive. Omitting any reference to Piaget's stages, and using the sloping lines merely as markers, the diagram would probably not cause offence to those who are unhappy about Piaget's stage theory.

It is necessary to comment on a number of features of the diagram. First, the discrepancy between the most able and least able widens with increase in age. Another way of looking at this is to point out that the difference in intellectual capability between the most able and the least able in a particular year group is considerable, particularly in the upper junior and secondary school. In comparison, the growth in intellectual capability from one year to the next for any individual child is small. Secondly, at the extremes of the ability range it is difficult to know what to do with the sloping lines which divide the stages. A few children, for example those with severe brain damage, make very little progress intellectually and do not fit into the implications of the scheme shown in Figure 5.3. Likewise, at the other extreme, a very few children seem capable of such rapid intellectual development that it is doubtful that the diagram accommodates them too. Thirdly, a considerable proportion of school-leavers of 16 have not reached the formal operational stage. Whether they ever do is not known since very little cognitive-development research evidence exists concerning such individuals. Taking into account the importance of motivation and intellectual activity in learning it seems very likely that a proportion of the population never develop those abilities outlined by Piaget as being characteristic of formal operational thinking. Finally, the diagram, in its simplicity, gives the impression of sudden change from one stage to the next. This, clearly, does not happen.

ACCELERATING LEARNING

From the preceding outline of Piagetian stage theory, itself only a part of the totality of Piaget's theoretical contribution to views on human learning, it might be thought that one implication is that there is little anyone can do to accelerate learning in others. This is not necessarily the case. Piaget's theory is not solely concerned with maturation; it is fundamentally about action and interaction. Even when engaging in abstract mathematics our thought processes are founded on previous action. Here there is a hint of the fundamental philosophical distinction between Piaget and the behaviourist tradition. To Piaget, it is not a case of pouring knowledge into an empty vessel. The acquisition of knowledge requires action on the part of the learner and requires interaction with the environment. Knowledge has to be constructed by each and every learner, hence different children must be expected to learn differently from the same learning experience. Interaction with the world outside the individual child carries the implication that an enriched environment could help to accelerate learning but only, of course, to the extent that the child can benefit through his or her constructive efforts.

Piaget's background as a biologist ensured that he regarded intellectual development in the same way as any aspect of growth, and in particular as involving self-regulation. When new ideas impinge on existing ideas it can happen that they create conflict. A situation of disequilibrium results, and this must be resolved. As a living being a child must reconcile any disturbance to the stability of his or her mental state. Piaget referred to this phenomenon under the description 'equilibration', and there are many who regard this aspect of Piagetian theory as the most important. Equilibration implies equilibrium in the same sense as in the natural sciences. It implies not a state of rest but a state of balance, a state for which the system is striving. In connection with equilibration Piaget introduced two helpful ideas, namely assimilation and accommodation. Assimilation refers to the taking in, the acceptance of, new ideas. Accommodation refers to what might be necessary in the way of modification and amendment to previously held ideas in order for assimilation to be possible. These two aspects of equilibration occur together and are generally inseparable. Equilibration, in the form of assimilation and accommodation, is relevant to all learning, but a few mathematical examples are appropriate.

We introduce young children to the idea of number. The children must master ideas of 'oneness', 'twoness', 'threeness' and so on, must comprehend the implications of the usual counting sequence 'one, two, three . . .', must absorb all relevant terminology and symbolism, must appreciate both cardinal and ordinal aspects, must learn how to apply the four rules, involving the use of place-value to organize our recording and manipulation of numbers, and all of this takes a long time. Over many years, children build up a view in their minds of what we mean by number and what a number is. This undoubtedly necessitates continual equilibration. Subsequently we introduce fractions. At the time of introduction there might be no suggestion that fractions are themselves also numbers, but eventually we hope that what is regarded by the child as a number is very much extended and modified from our original implication that 'numbers' = 'natural numbers' (possibly including zero). In assimilating the ideas that fractions are (rational) numbers, that improper fractions are still fractions and are numbers, that integers are numbers, that there are other (irrational) numbers, that the idea of real numbers is a valuable one, at each step the previously held view of what 'number' means requires modification. Assimilation cannot take place without accommodation, and accommodation might not be easy. In a trivial way we can appreciate this because in a puzzle situation, when asked to 'give a number between 1 and 10', the expectation is that the number required is a whole number or natural number. There is a strong tendency to think only of natural numbers when adults are requested to provide 'a number', so perhaps many of us have not accommodated fully, in the way that mathematics teachers might intend.

Another mathematical example involves equations. When equations are introduced they are inevitably linear equations. Techniques for solving linear equations are introduced and practised. The subsequent introduction of quadratic equations might well raise problems of accommodation. What we mean by 'equation' is certainly extended and methods appropriate for solving various kinds of equations need to be assimilated. The accommodation problem might be that techniques appropriate to linear equations no longer work; pupils certainly sometimes try to use them, however, showing that there are residual accommodation difficulties. Once a broader, more general, view of 'equation' is arrived at through the gradual introduction of a variety of different kinds of equation, there might be little problem when any further kinds of equation are introduced. Mastery of a broad view of equations as incorporating linear, simultaneous, quadratic and trigonometric, for example, should lead, for those who continue beyond this point, to minds which are open enough to accommodate, for example, logarithmic, exponential and differential equations.

At a higher level still, as was mentioned in Chapter 2, there is a widespread belief that velocity is proportional to force and also that force always acts in the direction of motion. These views arise through construction by the individual on the basis of observation of the world. They are, however, incorrect, and it took mankind many thousands of years to arrive at correct views. It should not be a surprise that such beliefs are prevalent, nor that they are very resistant to teaching. It is even possible to find individuals who show mastery of the mathematics (and physics) in the classroom but whose views outside the classroom revert to the popular incorrect notions. This is a situation where accommodation might even require complete eradication of previously held views. It even appears that the correct laws can be assimilated and live alongside incorrect laws. The individual must arrive at a state of mental equilibrium, even if that means one law for the classroom and another for the real world.

There are issues raised by the concept of equilibration which relate to the acceleration of learning. Is it possible, for example, to accelerate learning by setting out to avoid some occurrences of disequilibrium? Is it wrong to introduce pupils to the concept of a simple balance as a parallel to solving linear equations on the grounds that the comparison with balancing weights no longer works when we move on to quadratic equations and differential equations? Skemp (1971) has referred to the problems of using inappropriate schema which are not applicable beyond a certain type of mathematical situation. Nuffield (1969), in introducing integers, criticized such devices as temperature scales which took children some way into a study of integers but had to be rejected when it came to multiplication and division. Unfortunately, it is not easy to find devices or schema which are better than thermometers (in many countries) for introducing integers and balances for introducing equations. When these devices subsequently fail, however, disequilibrium may follow.

There is a contrary view, namely that by deliberately placing a child in a state of mental disequilibrium you force the constructive activity required for accommodation and learning is more permanent than if ideas are presented passively. Evidence from using equipment as a basis for experiment and discussion in mechanics learning suggests that the presentation of conflict which needs to be resolved can lead to successful learning. Given a loop-the-loop toy as an aid to studying motion in a vertical circle (see Figure 5.4) students have been observed to propose theories which

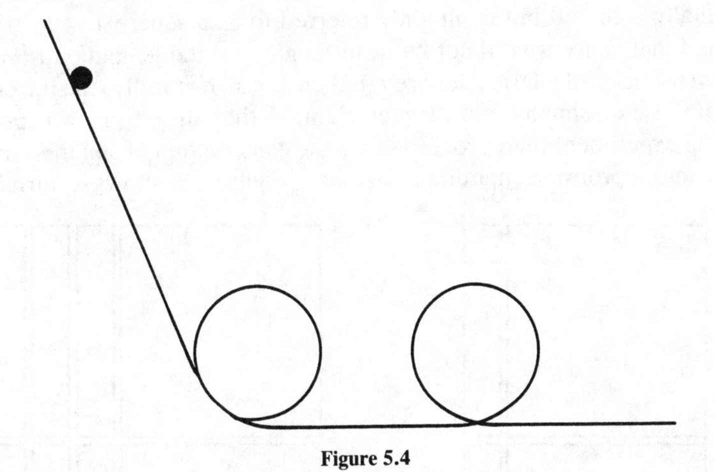

Figure 5.4

were incorrect but based on intuition which could be tested out and found wanting (Williams, 1985). The conflict between incorrect theory and observed results led to further theory and experiment and eventually to acceptance of the correct laws. Such learning is likely to be more permanent than any attempt to present the correct law without active involvement. The latter is likely to lead to one law for the classroom and another for outside. Examples of incorrect theories which might lead to useful and constructive conflict are: that the completion of a vertical circle depends on the steepness of the approach track, and that the object needs to be released from a point on the approach track level with the top of the circle in order to complete a circle.

The views of Bruner concerning learning are relevant to a consideration of the acceleration of learning. The following statement from Bruner (1960b) is well known: 'We begin with the hypothesis that any subject can be taught effectively in some intellectually honest form to any child at any stage of development.' This certainly appears, on first reading, to be a complete contradiction of any suggestion that particular topics may be assigned an absolute level of difficulty according to Piagetian stages of development. One experiment described by Bruner (1966) involved comparatively young children learning about quadratic expansions, not normally taught until at least age 13, by using some of the equipment suggested by Dienes (1960), illustrated in Figure 5.5.

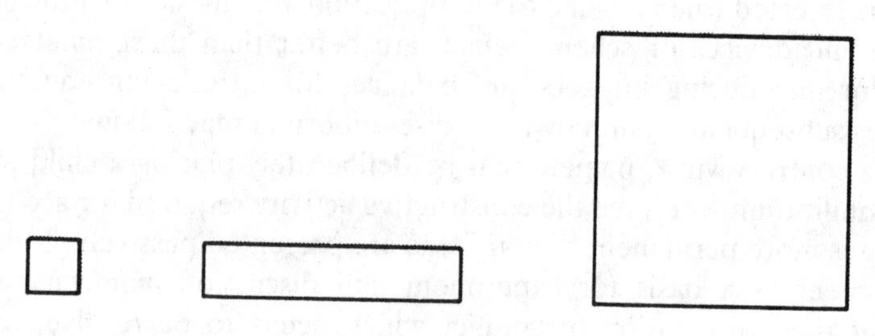

Figure 5.5

It was necessary to ensure that the children accepted the dimensions of the small square (actually a cuboid but commonly referred to as a square!) as 1×1 and that they accepted that, since we did not know its length, a suitable name for the strip was 'x'. The dimensions of the larger square are then $x \times x$. Naturally, the introduction of these notions is not simple, but Bruner claimed that they were accepted by the children. The experiment then proceeded via the construction of squares larger than $x \times x$ by putting appropriate materials together. Figure 5.6 shows a variety of such

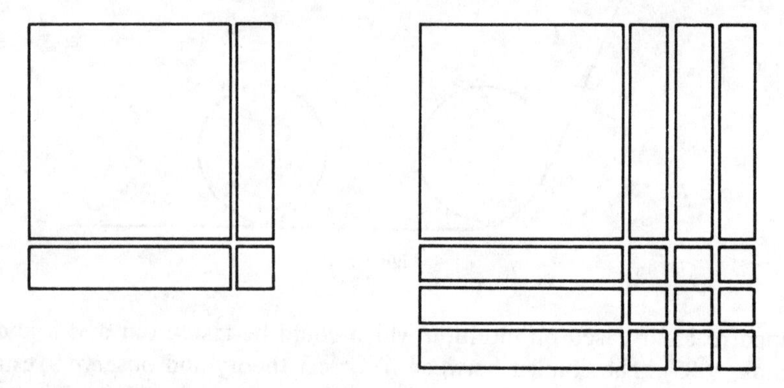

Figure 5.6

larger squares. The introduction of notation to record these results then became necessary, and this again was not simple, but was achieved through such steps as, 'one x-square plus two x-strips plus one unit square', to 'one x□ plus two x plus one', and finally to 'x□ + $2x + 1$', which could then be equated with $(x + 1)$□.

At all stages, because so many new ideas were being considered, particularly involving the use of notation, time for discussion and thought was important. At the end the children, it was claimed, had learned some mathematics several years in advance of what might have been expected. The ideas that

$$(x+1)^{\square} = x^{\square} + 2x + 1,$$
$$(x+2)^{\square} = x^{\square} + 4x + 4,$$
$$(x+3)^{\square} = x^{\square} + 6x + 9,$$
$$(x+4)^{\square} = x^{\square} + 8x + 16,$$

and the studies of number patterns which could emerge, leading to further generalization, are all very advanced for young children. The method can eventually lead to the practical demonstration of the use of the distributive law in more general cases like

$$(x+2)(2x+3) = 2x^2 + 7x + 6.$$

It is important to note that the children concerned had not been introduced to the ideas in the way which would be considered normal for, say, 15-year-old pupils. This would have been impossible if only because the symbolism would not have been available. However, in an intellectually honest way appropriate to their stage of development they had been introduced to a much more advanced mathematical idea than is normal. The question must arise as to whether the results of the experiment are in any way in conflict with the kind of conclusions about the acceleration of learning which one might draw from Piagetian theory. There was no suggestion in the results of the experiment that children had learned advanced mathematics in a formal way and were able to use the ideas and relate them to other aspects of formal mathematics. In stating that proportionality is an aspect of mathematics which requires formal operational thinking it is implied that we mean reasonable facility in the use of equality of ratios in arithmetic manipulative situations. The idea that because one measure is larger another must be comparably larger, would not be considered a sufficient indication that metric aspects of proportion were understood.

Certain other aspects of Bruner's experiment have led to criticism by mathematics teachers. In terms of practical applicability in the ordinary classroom there was no suggestion from Bruner that teachers might be able to use the procedure without assistance and with an entire class of 30 or more pupils. The original experiment reported by Bruner and Kenney (1965) involved only four children, with six adults available in the role of teachers. The four children were only 8 years old but were in the IQ range 120–130 and came from middle-class professional homes. The classroom environment was as favourable as is possible, in terms of the availability of concrete resource materials, suitable work assignments and, of course, opportunity for question and discussion with adults. One must accept the results for what they are—the results of one experiment which show what might happen under particularly favourable circumstances. One must be cautious about assigning absolute levels of difficulty to topics in the mathematics curriculum. There are almost certainly ways into the topics which are appropriate for young pupils which favourable circumstances might make possible.

Curriculum Bulletin No. 1 of the Schools Council (1965) recorded accounts of work carried out by 10-year-old pupils. In one particular instance one child was so motivated by a study of gradients that he pursued the idea to the beginnings of differentiation and integration. Other children became involved in what was an intellectually honest and justifiable approach to elementary calculus taking into account the stage of development of the children. It is, of course, possible that those pupils who took the lead in the investigations were intellectually amongst the most able in the class.

However, that does not contradict the view that the assignation of absolute levels of difficulty to particular mathematical topics is not easy and might be dangerous and unhelpful. The issue of readiness for learning is a very complex one. We should certainly not use general statements about stages of development to justify not looking for appropriate ways of helping children to learn mathematical ideas, just as we should not present mathematical ideas to pupils in such a way as to persuade them that mathematics is not for them.

In terms of readiness, stages of development and the accelerating of learning, the work of the van Hieles is important, as is also the considerable body of research carried out in many countries to test their original hypotheses. The van Hieles postulated sequential levels of geometrical thought but, in addition, suggested phases of instruction intended to enhance the learning of geometrical ideas. The earliest studies had indicated that there were times when it appeared that learning had stopped, and the teacher was unable to take pupils further until it seemed that the children had reached a higher level of thinking (see Fuys, Geddes and Tischler, 1988). At first, five levels were conjectured, and these may be briefly described as recognition or visualization, analysis, order, deduction and rigour. Subsequently, Pierre van Hiele has reduced these to three, basically the first two and a third one which encompasses all the last three of the original five. Subsequent research, largely carried out in the USA, has studied the basic thesis (that levels can be identified, are discrete and form a hierarchy), the pattern of levels in given populations, and the possibility of basing instruction and learning materials on the model. One interesting finding is that 'the discreteness and the globality of the levels are doubtful, meaning that a child can act at different levels in different contexts and can even change level within the same task' (Hershkowitz, 1990). Although this might seem reminiscent of the problem of décalage within Piagetian theory, the literature reveals genuine enthusiasm about the likelihood of finding ways of improving the learning of geometry by building on knowledge obtained from research into the van Hiele levels.

CURRICULUM IMPLEMENTATION

The work of Piaget has probably been more influential than has the work of any other theorist in terms of mathematics curriculum development in Britain, particularly at the primary level. This is in stark contrast to the situation in the USA where the behaviourist tradition was not seriously questioned before Bruner began to inject new ideas from cognitive psychology, some of which were influenced by the work of Piaget. One illustration of the impact in England and Wales is the Schools Council bulletin referred to in the previous section. Although this publication was principally a practical guide for teachers, it included a chapter on research into the way children learn. There were many references to Piaget in that chapter and the impact that the work of Piaget was having on those concerned with improving the primary mathematics curriculum at that time is illustrated by the following extracts.

> Piaget set himself the task of finding out . . . how the principles of conservation and of reversibility, as applied to numbers and to spatial thinking, develop in the minds of young children. The two principles are fundamental to all mathematical (and logical) thinking.

> . . . understanding cannot be taught nor does it come by itself, independently of experience . . . This does not mean that there is nothing the teacher can do except wait for the dawn of understanding. He can provide the kind of experience which will assist the child to move from intuitive to operational thinking.
>
> Children learn mathematical concepts more slowly than we realized. They learn by their own activities.
>
> Although children think and reason in different ways they all pass through certain stages depending on their chronological and mental ages and their experience.

The report of the Mathematical Association (1970) on primary mathematics included an appendix on 'Understanding and mathematics' which incorporated much of the spirit of Piaget within a broader review of what was known about learning. Caution was recommended, however, in the application of any interpretation of Piagetian theory, as is shown in these two extracts:

> Although these stages . . . have been broadly substantiated by a large number of research workers, we should show due caution in accepting them as a permanent feature in childhood development.
>
> . . . it is important not to discourage experiment, in the belief that what has been found is an unalterable feature of childhood development. We have only to compare the thinking of primitive adults with that of educated children in industrial societies to see the vast changes which are possible.

Piagetian views provided the underlying rationale for the book by Lovell (1971b) prepared as a guide for teachers of young children. The two extracts below would be found acceptable by many teachers today, even though certain aspects of Piagetian theory have received considerable criticism.

> There appears to be a danger that some mathematical ideas are introduced too early to children, or that there is insufficient appreciation on the part of professional mathematicians that many of the ideas they would like to introduce to elementary school pupils are understood only in an intuitive and not in an analytic sense by the children.
>
> It is not in any sense suggested that the child must always be 'ready' for a particular idea before the teacher introduces it. The job of the teacher is to use his professional skill and provide learning situations for the child which demand thinking skills just ahead of those . . . available to him . . . When a child is almost ready for an idea, the learning situation provided by the teacher may well 'precipitate' the child's understanding of that idea.

The distinction between 'ready' and 'almost ready' might be considered too subtle to be useful, but the message that the teacher does have a role to play and that it is not appropriate to sit back and wait is clear.

The early 1960s to the early 1970s was a time of curriculum experimentation in mathematics and a considerable number of curriculum projects sprang up. The only project which overtly expressed a Piagetian view was also the only major project aimed at younger pupils, and that was the Schools Council/Nuffield Project. The title of the first publication which introduced teachers to the work of the project was *I Do, and I Understand* (1967a). The words of the title themselves reflected the Piagetian message, though it was claimed that the title formed part of a Chinese proverb. This particular book included a discussion on how children learn and contained reference to stages of learning, to active learning and to the role of interaction with the environment. Amongst the many publications for teachers produced by Nuffield were three,

Checking Up (*I, II, III*), which provided teachers with Piagetian tests which could be applied to pupils in order to ascertain readiness in terms of the stage of development. The authors were at pains to point out that the tests were put on trial in schools before publication, as was the other material of the project, but the original ideas came from the work of Piaget. The *Checking Up* books also included a sort of time chart of conceptual development, particularly through the concrete operational stage, showing which ideas needed to be developed before other concepts could be mastered.

In *I Do, and I Understand* there were references to readiness, in relation to development through stages, and to the slow rate of intellectual development:

> Any attempt to hurry children through this stage of development (concrete operations) is liable to lead to a serious loss of confidence. They will discard real materials themselves at the appropriate moment . . . and eventually, when faced with a problem, will ignore all available materials and approach it abstractly.

Those who are familiar with the material of the Nuffield Project, however, cannot fail to have speculated on how a project which was apparently so heavily influenced by Piaget could include, for example, the approach to integers through ordered pairs. The reasons for doing this were explained, and a variety of practical or game-type activities were suggested to help in introducing this approach to integers, but they could not wholly compensate for the fact that, in the end, abstraction of quite a high level was required.

The Cockcroft Report (1982) contained many references to aspects of learning but showed no great enthusiasm for Piaget's stage theory or for any views on readiness which might be said to follow from Piaget. Understanding, and rate of intellectual development, are recurring themes, however, as reflected by these extracts:

> it is not possible to make any overall statement about the mathematical knowledge and understanding which children in general should be expected to possess at the end of the primary years.
>
> the curriculum provided for pupils needs to take into account the wide gap in understanding and skill which can exist between children of the same age.

The review of research, carried out for the Cockcroft Committee by Bell, Costello and Küchemann (1983), contained a section on research relating to stages of intellectual development and to the work of Piaget. This was an important section because it clearly revealed ways in which subsequent research to that of Piaget has raised doubts about the value of the idea of stages in relation to learning mathematics. Criticisms and current attitudes to aspects of the work of Piaget and to readiness are considered in the next section of this chapter.

CRITICAL EVALUATION

The work of Piaget has generally been welcomed as helpful in relation to curriculum design and to the planning of learning activities and experiences for children, but at the same time a body of criticism has grown up. Given the complexity of human learning and the comparative youth of educational psychology in the history of mankind it would have been very surprising if criticisms had not been levelled. Any major theory which appears to fit experimental data is likely to lead us forward towards widely acceptable theories about learning but may not provide the ultimate answer. In accepting that there have been criticisms of Piaget's work it is important to realize that many have been concerned with particular aspects, some relatively small. We must not necessarily reject every aspect of the theory because certain parts do not stand up to close inspection.

At a fairly trivial level there has been criticism of many of the tasks used by Piaget and on a number of grounds, listed below:

- many questions are not meaningful to the children—either they do not relate to the world in which the child lives or they do not motivate
- some questions might be regarded as ridiculous or frivolous for the above reason or because they contain questionable statements
- the complexity of instructions in some questions, the demands of pure language, are too much for some pupils
- some questions are not sufficiently free from context variables to produce results, from different backgrounds, which are comparable
- some questions, particularly those devised to test formal operational thinking skills, are too difficult even for most adults.

The following examples illustrate between them a number of these concerns.

One difficulty of task construction is to devise a question which is both meaningful and mathematically appropriate, and which is fair to every child. The task devised to expose a child's competence in handling ratio and proportion based on the eating habits of eels is an interesting case in point. In the Piagetian version (translation) eels eat meat balls and biscuits (Lovell, 1971a). In the Concepts in Secondary Mathematics and Science (CSMS) project version (Hart, 1981) the eels eat sprats and fish fingers. The whole situation, in both cases, is artificially contrived to set up a particular mathematical task. This is done frequently, but perhaps might be one reason (out of many?) why some pupils find mathematics unpalatable and divorced from the real world. It is suspected that children will play along with us under such circumstances, but does it affect the results? What of those children whose knowledge of biology is such that they cannot accept biscuits as part of the normal diet of eels and cannot accept that eels have mouths which are big enough to eat sprats? The particular stretching of the imagination required in this question is unimportant; the general principle is not. Are the results of research distorted because some pupils do not play along with us and generally do not respond to the question in the intended way?

Another Piagetian question concerns two fields each containing a cow and in which houses are to be built. In one field houses are placed randomly but in the other the houses are placed in a continuous line, in terraced format. At all times when questions were asked there were the same number of houses in each field. The basic question was whether each cow had the same amount of grass to eat, and this question was put for a variety of different numbers of houses. The results were as intriguing as those for the conservation tasks described earlier, but children did generally agree that the two cows had the same amount of grass to eat when there were no houses or just one house in each field. With more houses than one, younger children denied conservation. However, there is a complication which only certain children, particularly rural children, have been known to introduce. Cows have a tendency to churn up the ground near farm buildings, killing off the grass. Would this happen round the houses in the question? If it did there would be less grass to eat in one field than in the other! The context of the question is not equally fair to all children, so results might be distorted.

Yet another Piagetian task involved wooden beads in a box. Most of the beads were brown, but a small number were white. The question was whether the box contained more wooden beads or more brown beads. Results suggested that many children were unable to answer correctly. The question, however, is one that most adults would puzzle over, not because they could not answer the question but because they might not believe that the questioner had expressed it correctly. It is unusual, and to some extent ridiculous, to ask a question comparing the number of objects in a set with the number of objects in a subset of the set. The description of the application of the test by Piaget suggests that he was at great pains to try to ensure that the question was understood. But there is still a nagging feeling that the unusual and unexpected, even unacceptable, nature of the question might have seriously influenced the results.

The above examples illustrate how difficult it is to produce or invent a question which is acceptable from all points of view and upon which one can base general conclusions about intellectual development. There have been many experiments carried out to test Piaget's conclusions using alternative language or situations but based on the same ideas. Results from such experiments have thrown serious doubt on aspects of the conclusions drawn by Piaget. Yet it must be admitted that no alternative experiment has ever produced results in which all children across a wide age range have answered correctly. What has normally happened is that the proportion of children answering correctly has been different, occasionally very different.

Many of these issues have been discussed by Bryant. Another ability required for mathematical development not previously considered here and which Piaget had claimed was not present until the onset of concrete operations was the ability to make transitive inferences. A simple task involving transitivity might be based on three quantities A, B and C, direct comparison revealing that $A>B$ and $B>C$. Piaget concluded that pre-operational children could not deduce that $A>C$. Bryant's discussion of this issue (Bryant, 1974) is well worth studying. Such study reveals, first, the difficulty of devising a battery of test situations which avoid all criticisms of the types discussed above and which are acceptable from all points of view, and, secondly, the discrepancies between the results of the alternative experiments used and the Piagetian experiment. Bryant was particularly concerned that Piaget had not taken into account alternative reasons for the failure of young children to make transitive inferences. A very real alternative reason was considered to be that young children could not keep in mind the two earlier comparisons, $A>B$ and $B>C$, which is essential before inferencing becomes possible. Memory training involving such relationships did, indeed, produce a higher success rate in inferencing. Bryant also took into account the possibility that children might state the correct conclusion for the wrong reason, that is, because A is the larger in $A>B$ and because C is the smaller in $B>C$, any relationship involving only A and C must have A as the larger and C as the smaller. In conclusion Bryant stated,

> This experiment demonstrates conclusively that young children are capable of making genuine transitive inferences. Two main points follow from this conclusion. The first is that Piaget's theory about logical development must, to some extent, be wrong.

The Piagetian task involving beads and brown beads has been investigated by

McGarrigle and the conclusions are reported in Donaldson (1978). McGarrigle invented an experiment based on toy cows, three of them black and one white. When laid on their sides the cows were described as 'sleeping'. He was then able to compare the standard Piagetian form—'Are there more black cows or more cows?'—with a version which introduced greater emphasis on the total class—'Are there more black cows or more sleeping cows'. The cows were all sleeping in both versions of the experiment. With six-year-old children the success rate increased from 25 to 48 per cent with the introduction of the word 'sleeping'. This illustrates the kind of influence particular language can exert on test results. In another experiment black and white toy cows and horses were arranged on either side of a wall, as in Figure 5.7. The

Cows

B B W W

B B B W

Horses

Figure 5.7

children were asked a number of questions, including, 'Are there more cows or more black horses?' Only 14 per cent of children answered this correctly. Donaldson concluded from the results and the accompanying comments of the children that they were comparing the black horses with the black cows. Children's own interpretations of the language used in framing questions clearly can affect results and conclusions drawn from such results can be distorted. The issue of language in mathematical education is discussed more fully in Chapter 8.

Piaget has also been criticized for his lack of concern about sampling. Generally his experiments were tried out on very small numbers of children, seemingly often a few from the immediate locality. As a biologist, however, he may have been very content in working with a few readily available subjects. Biologists often have to be content with very small samples in experimental work. Conclusions drawn on the basis of work with small samples must be cautious, but the information which is itself obtained is not necessarily invalid. Many experiments based on Piagetian tasks have been replicated by others in many countries around the world using much larger samples including children spanning the whole range of all relevant variables like age, ability and social background. Because of this, the criticism of inadequate sampling is perhaps not one to which we need give much attention.

A much more substantial criticism of Piagetian theory concerns the idea that human intellectual development occurs in stages in which qualitatively different thinking structures may be detected. All critics draw attention to the weaknesses of such a theoretical position. There is just too much variation to make the theory useful, particularly in terms of prediction. There is considerable evidence that individual children cannot easily be categorized as being at a particular stage of development. For example, in one subject area they might reveal complete competence in tasks which are considered to require formal operational thinking, yet in another subject area they might reveal no higher than concrete operational responses. Even in the same subject area, say conservation, they might behave on some tasks as if they are at the concrete operational level but on other tasks their behaviour might indicate only pre-operational thinking. One day they might answer a particular task correctly, suggesting capabilities associated with a particular Piagetian level, but on the next day, on an exactly comparable task, their reaction might suggest they have not reached that level. Piagetian theory acknowledges this phenomenon in two ways. First, between any two adjacent stages there is a period of transition when such phenomena must be expected. Secondly, some element of confusion of the sort described above, referred to by Piaget as 'décalage', is accepted because humans will always be prone to respond with a degree of variability. Many modern critics, however, are unhappy that the variability seems too great to support a theory which is of any real use. They would not necessarily reject any suggestion that intelligence or cognition develops throughout childhood, that maturation of the central nervous system plays a part and that the quality of interaction with the environment is an important contributory factor. They are unhappy with the idea of development in identifiable stages.

Another criticism arises in the accusation that Piaget's theory is only illustrative and not confirmatory (Brown and Desforges, 1977). The data may be considered to fit the theory but there is no other way to prove the theory. Certainly, any theory which can only be supported in this illustrative way cannot be regarded with satisfaction, but it is likely that human progress in many areas of science has depended on such assumptions. It would be interesting to research into the history of science with a view to establishing how many theories were originally developed because they fitted the facts. More abstract proofs can only be sought once the data has suggested a hypothesis. In education, we might not yet even be sufficiently certain of our data.

Piaget's work has also been taken to imply a consistent order in the acquisition of mathematical concepts. The time chart contained in the Nuffield *Checking Up* books has already been mentioned. Cross-cultural studies, however, have generally not confirmed consistency of order. It is possible that there are elements of order, that length must come before area, for example, but there are also many indications that in different cultures the order is not the same as in Western culture. Piaget's theory does acknowledge the vital effect of interaction with the environment in its widest definition. The role of experience is much more crucial than any constant order theory can acknowledge. It is, for example, often suggested that any differences in spatial and mechanical ability between boys and girls is because of differences in the environment in which the two sexes are reared. This is an hypothesis which cannot easily be proved or disproved but it cannot be denied that this is an example of different environments within the same overall culture. Children from more rural and agricultural cultures tend to develop skills and knowledge and understanding required for survival and not develop other capabilities which those in an industrial society might expect.

All of the above comments and criticisms of aspects of Piagetian theory should suggest that attempts to apply Piaget either to the construction of the mathematics curriculum or to assessing pupil progress have not been successful. On the whole this has been the case. The only mathematics curriculum development project in Britain which has referred heavily to Piaget was the Nuffield 5–13 Project. This project in itself received rather limited and disappointing levels of support and interest from teachers, who tend naturally to be more interested in the practicalities of schemes of work and exercises for pupils rather than directly on the implications of Piaget for classroom implementation. Nuffield was comparatively lacking in material of direct applicability to the classroom. More recently a new scheme entitled 'Nuffield Maths 5–11' has appeared as 'a revised and extended version of the Nuffield Mathematics

Teaching Project'. The revision has been such that anything that smacks of pure Piaget has been removed. The nearest message one can find to that of the original Nuffield in terms of the influence of Piaget is perhaps that:

a) Children learn at different rates and so will not reach the same stage simultaneously;
b) Young children learn by doing and by discussion;

and the message about exposition, which some would regard as being no part of the Piagetian message, is also included in

c) As well as finding out and 'discovering' things about mathematics, children need to be *told* things about mathematics, particularly if new vocabulary is involved.

(Nuffield Maths 6, 1983)

Difficulties in relating to the Piagetian stage theory have been experienced by those who have researched into mathematical understanding on a large scale. Hughes, writing about the Schools Council Project on the development of scientific and mathematical concepts in children between the ages of seven and 11 (Hughes, 1980), stated that 'The conclusions . . . confirm the doubts one has for the resolving power of Piagetian type tests.' The study was based on the responses of 1000 children to a battery of practical tests on the concepts associated with area, weight and volume. It developed 'as a result of the trend in the late sixties and early seventies for modes of teaching to be loosely based on beliefs about the conceptual development of children (basically Piagetian)'. The results exposed so much décalage as to make general statements about conceptual development, particularly of stages of intellectual development, completely inappropriate. The following extract sums this up:

> Some children, at all ages, grasp one conservation concept in one test situation before grasping it in another; this is true from topic to topic and also between fairly similar tasks in any one topic (say weight). From our research it is not possible to determine for certain which they will grasp first.

Reference to selected results of the work of the mathematics team of the CSMS project has already been made (Hart, 1981). This team, too, experienced difficulty in relating their findings to Piagetian theory. It had been hoped that the development of the understanding of mathematical concept areas 'could be described in terms of the demand (as related to) Piagetian levels of cognition', but this did not materialize. The problem was clearly stated:

> It was hoped that a child could be designated as being at a particular Piagetian level and, by looking at his performance on a maths test, the mathematical levels could also be described in Piagetian terms . . . We found, however, that the child's performance varied considerably task to task and that we could not label a child as being overall at a certain Piagetian level. (Hart, 1980)

Nevertheless, the terminology 'concrete operations', 'formal operations' and the like was, and still is, apparently found to be useful by those reporting on empirical research, and is yet in use by those who have subsequently proposed developmental as opposed to behaviourist theories of learning, and who will be considered in subsequent chapters.

More recent work by Booth (1984) reflects the dilemma facing empirical researchers in relation to Piagetian theory. Reporting on the research project 'Strategies and Errors in Secondary Mathematics: Algebra' she draws attention to inconsistencies militating against 'the unqualified acceptance of the "unified" stage view of cognition which characterizes the Piagetian formulation'. At the same time the following points are also made:

> the observed similarities in the nature of the informal methods used by different children, as well as the points concerning context and the generalized nature of algebraic representation outlined above, suggest some generality in cognition which requires explanation

and

> Analysis of the *nature* of the difficulties which children have been observed to experience suggests a picture of conceptual growth which is generally not inconsistent with Piaget's description of the development from concrete to formal operational thinking.

Despite criticism, expected in the world of education, Piaget has acted as an inspiration to many who have produced alternative theories concerning cognitive development, for example Ausubel, and also to those who have proposed ideas concerning mathematics-learning, for example, Dienes. Shulman wrote (1970), 'Many psychologists are seriously suggesting that his [Piaget's] stature will eventually equal that of Freud as a pioneering giant in the behavioral sciences.' The influence of Piaget will certainly reappear in subsequent chapters of this book.

CROSS-CULTURAL ISSUES

The implication from the work of Piaget is that all children, all over the world, develop in substantially the same way. To Piaget, progression from one stage to another is achieved through a combination of maturation, interaction with the environment and equilibration, with equilibration being perhaps the most critical factor. But what is the comparative effect of these various factors? Might it be that the wider environment, which includes the social and cultural environment, plays a larger part than Piaget believed? It must be admitted that Piaget's theoretical constructs were produced within a Western cultural situation and on the basis of empirical work carried out with children brought up in such an environment. Berry (1985) has summed up the criticisms of the assertion that development is invariant across different cultural and linguistic groups in writing 'studies conducted in a variety of settings in Latin America, Africa, Asia and the South Pacific have raised serious questions about the validity of this assumption'.

Many cross-cultural studies have, in fact, been carried out, but many of them by Western research workers, who Piaget himself pointed out would find this kind of research very difficult, because such studies presuppose a complete and thorough knowledge of the language and a sophisticated understanding of the culture and society within which the research is to be conducted. The same point could, of course, be made about research with children from particular subcultures within Western society. Dasen (1972) reported that 'A number of studies show that both non-Western and low socio-economic class Western children lag behind in their concept development when compared with middle-class Western children'. Indeed, 'since Western science does not necessarily represent the form of thought valued in other cultures, nor in fact in some subcultures within the West, the Piagetian sequence is likely to be ethnocentric' (Dasen, 1977). A particular issue is the one of language, namely whether the language used to communicate ideas allows the complete elaboration of concepts which have largely been derived through the medium of Western languages, and whether this communication is conducted in the first or a second language. The issue of language is discussed in more detail in Chapter 8.

A major difficulty in making sense of the findings from cross-cultural studies has been the absence of adequate common ground, with each new study being perhaps in a new culture and with new tasks. Many studies, however, have deduced a 'time-lag' in the development of concepts. Such studies should not be interpreted in terms of cultural deprivation, with the development of the Western child being taken as the norm against which all other children must be compared. Concepts might develop together in a Western society but at different rates in another culture when that culture places a different value on the ideas or when one idea is more culturally relevant than another. The early study by Gay and Cole (1967) established that for the Kpelle, a group in Liberia, it was the requirements of everyday life which dictated which numerical estimation skills were most fully developed, and that with certain skills the Kpelle were better than American college students. Not all cross-cultural studies have adopted a strictly Piagetian framework, of course, but the results often throw light on the issue of social environment. Saxe and Posner (1983) have reported research carried out in the Ivory Coast which reveals that the children of merchants adopted more economical strategies of arithmetic problem-solving than children brought up in an agricultural situation, showing 'how individuals develop the symbolic skills that are most useful to them in their differing social contexts'. There is some support, across many different cultural groups, for believing that concepts of number conservation develop similarly through the concrete operational stage, yet time-lags are still widely reported and there seems to be considerable variability in performance across tasks. Saxe and Posner also stated that 'most studies inspired by the Piagetian approach succeed in documenting a shift from pre-operational to concrete operational mathematical concepts', but at the same time 'problematic for the Piagetian formulation is the lack of empirical support for the construct of stage'. The conclusion drawn by Saxe and Posner, based on evidence from a wide range of cross-cultural studies, is certainly that 'the formation of mathematical concepts is a developmental process simultaneously rooted in the constructive activities of the individual and in social life'. Without the obvious benefit of cross-cultural studies, Solomon (1989) has recently similarly criticized Piaget's assumption of a solitary knower who must construct mathematical understanding, suggesting that it is rather the case of 'an essentially social being for whom knowing number involves entering into the social practices of its use'. Such social practices will inevitably vary according to cultural background.

In terms of readiness for learning, the underlying theme of this chapter, the issue remains somewhat unresolved. The view of behaviourists is found to be unsatisfactory by many teachers, yet the presence of prerequisite knowledge, suggested by Gagné and supported by the cognitive psychologist Ausubel, is clearly important. There might well be a maturational factor in learning mathematics. The role of interaction with the environment is clearly important, and that environment comprises not only the immediate classroom or teaching environment but also extends to the wider social and cultural situation within which children are educated. To a certain extent teachers are forced to keep an open mind on the readiness issue. It would possibly be detrimental to the cognitive development of pupils to assume too quickly and easily that they are not yet ready for a new idea. But experience of teaching would suggest that attempts to introduce new ideas will not always be successful, and we ourselves must be ready for that.

SUGGESTIONS FOR FURTHER READING

Bryant, P. (1974) *Perception and Understanding in Young Children*. London: Methuen.
Copeland, R. W. (1979) *How Children Learn Mathematics*. New York: Macmillan.
Dasen, P. R. (ed.) (1977), *Piagetian Psychology: Cross-Cultural Contributions*. New York: Gardner Press.
Donaldson, M. (1978) *Children's Minds*. Glasgow: Fontana/Collins.
Lovell, K. (1971) *The Growth of Understanding in Mathematics: Kindergarten Through Grade Three*. New York: Holt, Rinehart & Winston.
Rosskopf, M. F., Steffé, L. P. and Taback, S. (eds) (1971) *Piagetian Cognitive-development Research and Mathematical Education*. Reston, VA: National Council of Teachers of Mathematics.

QUESTIONS FOR DISCUSSION

1 Is there more to readiness than the presence of certain relevant subordinate intellectual skills? Give reasons for your answer.
2 What do you understand by the terms 'concrete' and 'formal' and what are the implications for mathematics-learning?
3 When are children ready to learn about fractions (rational numbers) and operations on them?
4 For any topic which you teach regularly, state what you need to take into account in deciding whether a pupil is ready to learn the topic.

Chapter 6

Can Pupils Discover Mathematics for Themselves?

LEARNING BY DISCOVERY

All polygons with more than three sides (edges) have diagonals. A quadrilateral has two diagonals, a pentagon has five, a hexagon has nine, and so on (see Figure 6.1). A

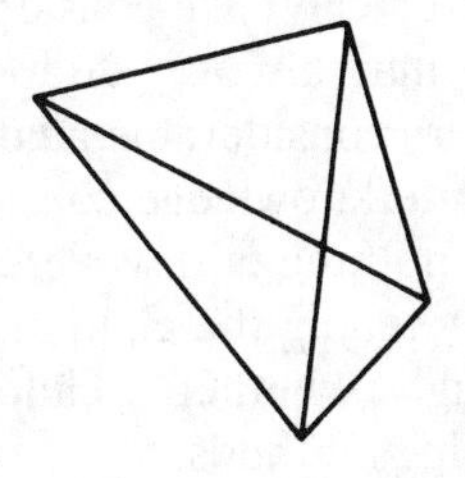
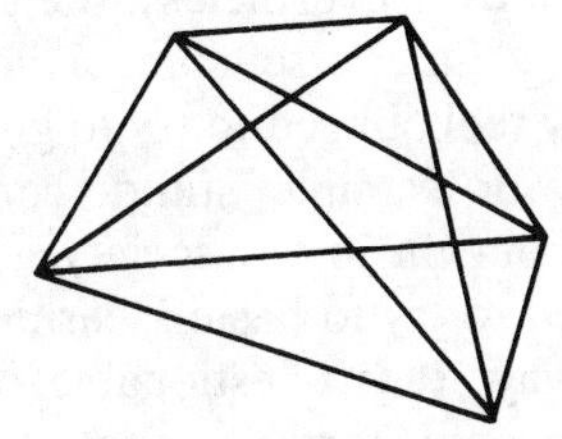
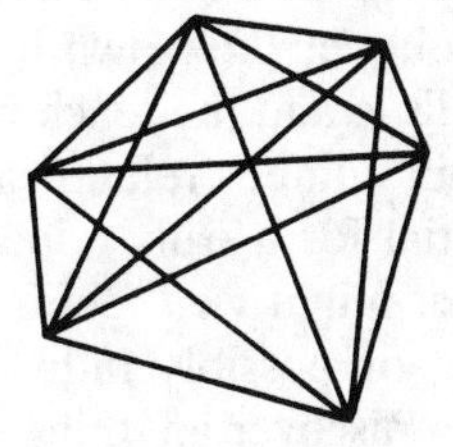

Figure 6.1

Number of sides	3	4	5	6	7	8	9	10
Number of diagonals	0	2	5	9	14	20	27	35

table of number of diagonals against number of sides reveals a number pattern which may be used to extend the sequence and enable the determination of the number of diagonals for any given number of sides. It is also possible to use the pattern to obtain a formula for the number of diagonals in terms of the number of sides. Extension activities might include investigating the number of regions into which the diagonals dissect the polygon when (a) the polygon is regular, and (b) the polygon is irregular. The entire assignment may be given to children to work on for themselves. Any results obtained would not constitute knowledge which all truly educated people must possess. It is unlikely that such knowledge would ever be taught in an expository way. The assignment would not be given to children unless the teacher believed or was persuaded that taking part in an active way in a mathematical investigation, and perhaps discovering some mathematics, was worthwhile.

Some teachers of young children regularly use coloured rods to help with the introduction to the counting of (natural) numbers. It is sometimes claimed, by advocates of the use of such apparatus, that simply playing with the rods takes the children a very long way towards mastery of number relationships (see Figure 6.2). The children discover that a certain pair of different coloured rods arranged in a 'train' are equivalent in some way (length) to a third rod of yet another colour. They might also ultimately discover that the rods may be arranged in a staircase of equal steps.

Figure 6.2

It appears that there is an extent to which mathematics may be discovered, and learning, it might be claimed, is the more thorough and complete for having been obtained by this means rather than by exposition. However, there is also an extent to which the teacher, or some other person, may need to intervene in order to introduce first the appropriate language, then to help clarify the thinking, and then to introduce symbolism and recording methods. Nevertheless, the children can have considerable control over their own learning. Of course, some children may discover disappointingly little and the teacher may feel obliged to try to give very considerable guidance. Certain number relationships, for example, still do constitute knowledge considered essential for a truly educated person in our society, so if nothing is discovered the teacher might well feel inclined to try to hasten learning by telling the child in some direct, or possibly indirect, way. It is questionable, though, whether a child who cannot discover anything can benefit from expository teaching methods.

Words such as 'activity', 'discovery', 'investigation' and 'problem-solving' have become very much a part of the language we now use in talking about mathematics teaching and how we should educate our pupils in mathematics. Many pupils today, however, are still taught largely by exposition and are given little opportunity to learn by discovery. Most teachers, when they themselves were pupils, were given almost no chance to discover mathematics, though some educators in every generation have believed that exposition alone was unlikely to prove efficient, particularly with younger pupils. With more able and older pupils teachers of mathematics have, in the past, often been able to avoid criticism with only minimal use of methods other than exposition and practice of skills. At the present time, however, there is much more pressure on teachers to use more active approaches. Supporters of the use of discovery, investigations and problem-solving have probably never been absent from the educational scene, but paragraph 243 and associated elaboration in the Cockcroft Report (1982) have given a boost to their cause. There is a certain vagueness about terms such as 'discovery', 'investigation' and 'problem-solving' which really does not matter all that much. It is the spirit of active rather than passive education which is at the heart of this issue and which is now considered important, not the exact definition of certain words which we use.

Shulman (1970) wrote of the new psychology of learning mathematics which, to a large extent, was based on discovery learning. The main advocate of discovery learning in the USA around and before 1970 was Bruner. Shulman described the origins of a theory of learning by discovery as a 'mélange of Piaget and Plato'. It was the work of Piaget, which has been interpreted in mathematical education as implying active interaction with the environment enabling the construction of knowledge and understanding by the individual, which was a major factor in justifying a discovery approach. In the USA this was a relatively revolutionary idea, given the previous domination of educational practice by behaviourist learning theory. It is interesting to speculate, however, whether behaviourism need necessarily rule out discovery, and indeed the idea of programmed discovery will be returned to shortly. Bruner's pioneering work in encouraging discovery learning in the USA was certainly significant for a considerable time and developments in Britain also reflected the same interest in more active approaches.

Curriculum Bulletin No. 1 of the Schools Council (1965) contained a number of references to learning by discovery at the primary level, for example:

> Mathematics is a discovery of relationships and the expression of the relationships in symbolic (or abstract) form. This is no static definition, but implies action on the part of the learner, of whatever age and whatever ability. It is the fact that mathematical relationships can be discovered and communicated in such a variety of ways that puts mathematics within reach of children and adults of all abilities.

A central message of the bulletin was that teachers must teach primary mathematics by means of as much active involvement as possible, using practical activities with equipment whenever possible, and by this means children would discover and would not need to be told. The principal author of this bulletin, Edith Biggs, has also written separately about discovery (Biggs, 1972). One interesting feature of Biggs' paper was the use of 'discovery', 'investigation' and 'active learning' in an almost synonymous way, reflecting an earlier point of this chapter. Another feature was her claim that discovery (investigational, active) methods gave pupils the opportunity to think for themselves and that only in that way could pupils realize their full potential. In addition, such methods generated real excitement for mathematics which, given the binding relationship between cognitive and affective factors in learning, no doubt contributed to the realization of full potential.

Different authors have attempted to classify discovery methods. The five described by Biggs (1972) provide a good means of reflecting on this issue, and they were: fortuitous; free and exploratory; guided; directed; and programmed. At one extreme, fortuitous discovery cannot be planned. It certainly happens but no learning programme can make use of it. At the other extreme, programmed discovery has the air of contradiction about it which has already been mentioned. The intention in programming a unit for work is to ensure that, as far as is possible, learning does take place. The unit 'Fibonacci fractions' is an example of an attempt of this kind.

This unit of work should provide an example of a discovery programme at the level of the reader. Successful completion of the unit cannot be guaranteed with any particular pupil or student, but successful completion does imply discovery of some mathematics, assuming the knowledge was not already held! The learning sequence is too rigidly sequenced to be guided. It is heavily directed and could be considered to be

Fibonacci fractions

You have already met the Fibonacci sequence

$$1, 1, 2, 3, 5, 8, 13, 21, 34, 55, 89, 144, \ldots$$

The numbers in this sequence may be used to form Fibonacci fractions, for example,

$$\frac{1}{1}, \frac{1}{2}, \frac{2}{3}, \frac{3}{5}, \frac{5}{8} \ldots$$

(1) Write down the next ten fractions in this sequence.
(2) Use your calculator to convert all fifteen terms into decimals.
(3) Draw a graph to show the value of the decimal for each term of the sequence.
(4) Describe, as fully as you can, what you notice about your sequence of decimals and, in particular, what would happen if you went on to 20, 50, 100 or more terms.

Now look at the alternative sequence of fractions,

$$\frac{1}{1}, \frac{2}{1}, \frac{3}{2}, \frac{5}{3}, \frac{8}{5} \ldots$$

(5) Write down the next ten fractions in this sequence.
(6) Use your calculator to convert all fifteen terms into decimals.
(7) Describe, with the aid of a graph if necessary, what you notice about this sequence of decimals.
(8) What is the relationship between the fifteenth terms of the two sequences? What do you think is the relationship between the limits of the two sequences? (There are two relationships, one based on difference and the other involving reciprocals.)
(9) Use your two relationships to write down a quadratic equation, the solution of which is the exact value of the limit of one of the sequences.
(10) Solve the quadratic equation to find the exact value of this limit and then determine the limit of the other sequence.

virtually programmed, though it might not pass a Skinnerian test of what constitutes programmed learning. Unfortunately, the lack of activity implicit in the unit renders it boring to some students. Others find it interesting and revealing. Allowing learners to work in pairs and discuss might alleviate the potential boredom problem.

Near the other extreme, discovery which is free and exploratory might follow from the investigation 'Rectangles'.

Rectangles

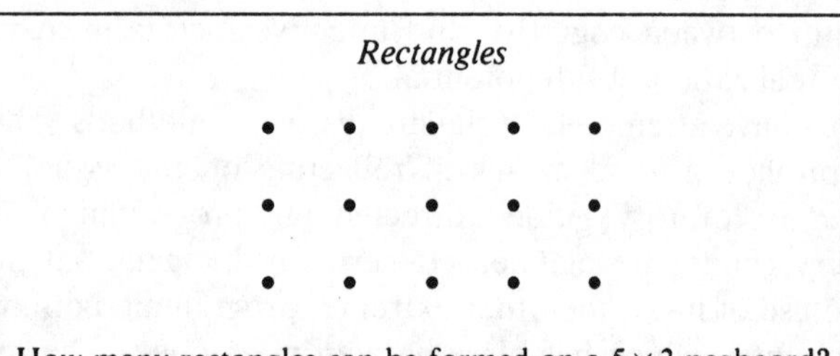

How many rectangles can be formed on a 5 × 3 pegboard?
What about other sizes of pegboard?

There is no guarantee that anything other than numerical information will be discovered from this investigation. What is more, the numerical information which is obtained might not be correct. Further, there is much more of a burden for the teacher if it is important that correct conclusions do result from free and exploratory activity. For this reason, with a firm eye on the syllabus and forthcoming examinations, teachers might be more happy to allow free and exploratory investigation if the results do not constitute essential knowledge. Where mathematical results are important many teachers would undoubtedly support teaching methods which they believe guarantee that pupils gain the requisite knowledge. Despite the missionary work of a few, teachers generally favour exposition under such circumstances. Biggs, and other advocates of investigational and discovery learning, have been at pains to point out that examination syllabuses can be completed and all required results learned thoroughly via active learning methods, but the majority of mathematics teachers have yet to accept this view. One course which appeared to attempt to integrate active learning and yet also teach a body of knowledge successfully was that by Bell, Wigley and Rooke (1978–9).

The value of discovery has been the subject of debate and some disagreement amongst educational psychologists. Gagné and Brown (1961) claimed to have established that guided discovery was the best method (of those used) to promote the learning of certain rules. There is not much evidence from research to support any particular view about the value of discovery methods, and we must not be swayed by the results of one experiment. Ausubel (1963) argued that guided discovery only looked best because of what it had been compared with—usually rote learning. He went further and claimed that there was just no evidence that discovery of any kind was a more effective teaching method than meaningful exposition. Ausubel did, however, agree that discovery is important in promoting learning with young children, and both Gagné and Ausubel agreed that active learning methods are more important for younger pupils than for older. Yet guided discovery is quite popular with some teachers. They believe the pupils are better motivated by an active approach, and perhaps by a challenge, but the teacher may justifiably step in at any time to ensure that the desired end-point is reached.

The enthusiasm for discovery learning generated by Bruner led to a debate with Ausubel, and both their viewpoints are important. The main points recorded by Bruner (1960a) in favour of learning by discovery were as follows. First, discovery encouraged a way of learning mathematics by doing mathematics, and encouraged the development of a view that mathematics was a process rather than a finished product. Secondly, discovery was seen as intrinsically rewarding for pupils, so that teachers using discovery methods should have little need to use extrinsic forms of reward. These two points carry great weight. Practical difficulties were acknowledged, namely that one could not wait for ever for pupils to discover, that the curriculum could not be completely open, so discovery would need to be to some extent guided or directed. Some pupils might even find their inability to discover extremely discouraging. It was, of course, up to the teacher to make the kind of judgements necessary to circumvent these difficulties. Such practical difficulties did not invalidate the case for active learning. The effort of trying to use discovery methods was worthwhile for what was achieved.

Ausubel (1963) attempted to temper this missionary zeal of Bruner rather than to try to dispute the validity of the main points because he felt teachers were likely to make excessive or inappropriate use of discovery. In the first place he suggested that discovery was not the only way a teacher could generate motivation, self-confidence and a desire to learn. Expository teaching, at its best, was just as capable of exciting and inspiring pupils. Not all reception learning was bad and not all discovery learning was good. Discovery could seriously de-motivate when nothing was discovered. Further, any suggestion that discovery learning implied creativity was questionable, for pupils can rarely be genuinely creative, and guided discovery was hardly creative. No research evidence was available which conclusively proved that discovery learning was superior to expository learning in terms of long-term learning gains. Certainly there was need for discovery methods with young children but discovery was not at all valuable for most learning at the abstract stage of cognitive development. Discovery, after all, *did* use up too much time. Practising mathematics as a process was not the main priority for school learning. It was much more important, for example, that pupils should learn the substantial body of knowledge which was essential for survival in a complex society. Since there was no possibility that pupils could re-create the whole of that knowledge, teacher intervention in a more or less direct way was necessary. Such were the views of Ausubel.

Discovery learning was adopted by some of those involved in curriculum development in mathematics in the 1960s and 1970s. Discovery was an important feature of the Madison Project in the USA. Davis (1966) drew attention to the place and value of discovery largely through examples of pupils' discoveries. The Madison Project claimed to use a discovery technique which they described as 'torpedoing', within more orthodox general discovery methods. The idea of torpedoing was that, once pupils thought that they had discovered a pattern, relationship or rule, an example was injected which did not fit, causing the pupils to think again. In a sense, this is an example of deliberately creating a state of mental disequilibrium, in order to encourage the twin processes of assimilation and accommodation, a tactic referred to in the previous chapter. It is not clear, however, that 'torpedoing' was sufficiently successful in promoting learning for it to be seriously advocated as a worthwhile technique.

In Britain, discovery methods were generally actively encouraged at the primary level through the work of Biggs and also through the Nuffield Project. Chapter 5 of *I Do, and I Understand* (Nuffield, 1967a) described the meaning and importance of discovery learning within the project. Many teachers will remember the early stages of secondary curriculum reform in the 1960s for the emphasis on changes of content. In the first report of the Midlands Mathematical Experiment (1964), however, we find, 'We are continually being surprised by what children *can* do, provided that it grows out of their peculiar experiences. Our job is to recognize mathematics in the children's activities and utilize it.' From 1968 onwards, the A–H series of the *School Mathematics Project* included experimental and investigational sections. More recently, in the wake of the Cockcroft Report (1982), there have been developments aimed at ensuring that secondary school mathematics curricula do involve an element of more active learning.

The efficacy, or otherwise, of discovery methods is, however, still under debate. The issue has been commented on by Davis (1984), as follows:

> one *cannot* compare, say, 'discovery teaching' with 'non-discovery teaching' . . . one can only compare *some specific attempts to do 'discovery' teaching*, vs. *some specific attempts to do 'non-discovery' teaching*. One or both may be done very well, or moderately well, or badly, or even very badly . . . One has NOT compared 'discovery' and 'non-discovery' teaching *in general. But that is the way the results are invariably interpreted.*

In this respect, discovery is no different from the subject of much other educational research. Supporters of discovery learning may therefore, to a large extent, be accepting a belief, summed up by Biggs (1972):

> I believe this method [discovery] is the best way to give our pupils real excitement in mathematics. I believe too, that it is only when we give our children a chance to think for themselves that they realise their full potential.

Research, in any case, usually attempts to measure only the quality of cognitive development, or what has been mastered. Gains in attitude to mathematics and increased awareness of the nature of the subject are not easily measured. Who knows what long-term benefits might accrue if discovery were to be used much more than it is, particularly at the secondary level where the balance, up to now, has been very much in favour of passive methods of learning mathematics?

GESTALT PSYCHOLOGY

Discovery learning depends on a child making connections and seeing relationships without having to have them pointed out by the teacher. Consider the quarter-circle problem. In order to solve this problem the child must see through the information to the whole structure of the situation and realize that the length of XY is equal to the length of the other diagonal of the rectangle, which is a radius of the circle. Such insight is frequently required in problem-solving, mathematical or otherwise.

A QUARTER-CIRCLE PROBLEM

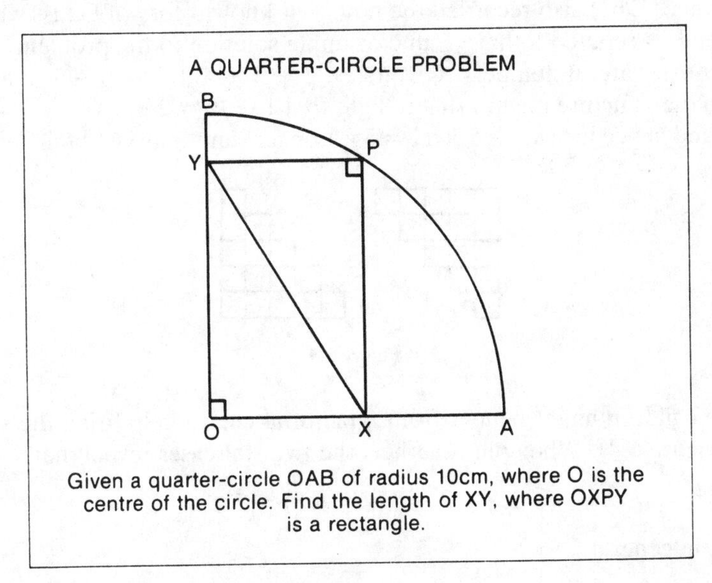

Given a quarter-circle OAB of radius 10cm, where O is the centre of the circle. Find the length of XY, where OXPY is a rectangle.

Insight, as a phenomenon in human learning, was acknowledged by Gestalt psychologists as being important. The essence of Gestalt psychology was, and is, that the mind (and not necessarily just in humans) attempts to interpret incoming sensations and experiences as an organized whole and not as a collection of separate units of data. If the underlying structure is immediately perceived in a meaningful way the learner is able to proceed with the solution to the problem. We, as teachers, can help our pupils to learn by providing experiences in which the structure is evident or by guiding or directing pupils to the structure. Gestalt psychology originally developed in Germany, and the word 'Gestalt', roughly translated, means 'form' or 'shape'.

The leading mathematical Gestalt psychologist, throughout the period of the development of the theory, was Wertheimer. Scheerer (1963), commenting on Gestalt psychology, reported on Wertheimer's parallelogram example as follows:

> Suppose a child who already knows how to get the area of a rectangle is asked to find the formula for the area of a parallelogram. If a child thinks about it, Wertheimer said, he will be struck by the fact that a parallelogram would look like a rectangle were it not for the fact that one side has a 'protuberance' and the other side has a 'gap' [see Figure 6.3] . . . Then he realizes that the protuberance is equivalent to the gap . . . Hence the formula is the same as it is for a rectangle.

Figure 6.3

Although trial and error might be involved in solving problems, where a problem has a structure of its own the structure helps point the way to a solution.

Wertheimer (1961) also recorded the now well-known story of Gauss who, as quite a young child, is reputed to have found a simple solution to the problem of summing the consecutive natural numbers. Given $1 + 2 + 3 + 4 + 5 + 6 + 7 + 8 + 9 + 10$, insight into the structure might bring to light that $1 + 10 = 2 + 9 = 3 + 8 = 4 + 7 = 5 + 6 = 11$, and hence the sum is $5 \times 11 = 55$. Longer sums may be obtained in a similar

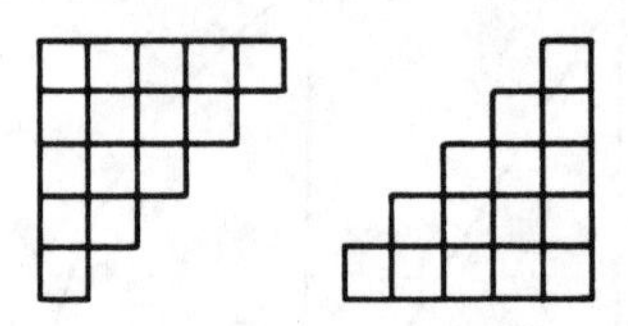

Figure 6.4

way. Some children might benefit from a pictorial cue to help bring the structure to light (see Figure 6.4). When put together, the two staircases reveal that

$$2(1 + 2 + 3 + 4 + 5) = 5 \times 6$$

and hence, in general,

$$2(1 + 2 + \ldots + n) = n(n + 1).$$

What is clearly suggested by Gestalt theory is that demonstration of a result by the teacher might not lead to insight for the pupil. Exposition of how to calculate the area of a parallelogram, perhaps based on proving congruence of the two small triangles, will not necessarily ensure that the pupils understand why it is required that the triangles should be proved congruent. Insight comes as an aspect of the discovery process. The situation needs to be structured so as to make the necessary discovery as certain as possible. The insight gained may then be transferred, and areas of triangles and trapeziums understood.

Gestalt theory was accepted and acted upon by Stern through the provision of structural apparatus. Stern, in fact, dedicated her book (1953) to Wertheimer. The Stern structural apparatus was devised to enable children to discover arithmetic for themselves through the prompting of insight which the equipment fostered. It was not enough to learn number by counting; the relationships between numbers needed to be made explicit. Thus, coloured rods were used as the basis, and these rods were segmented to enable pupils to see how many 'unit' rods were equivalent to a particular rod, and to see the difference in value of two particular rods (Figure 6.5). Since the

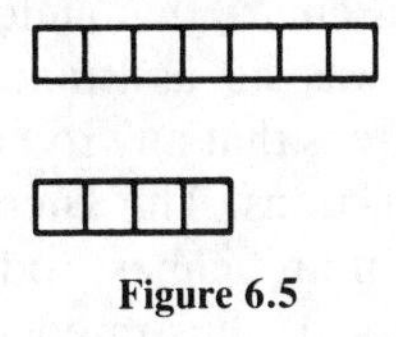

Figure 6.5

invention of the Stern apparatus many other forms of structural equipment have been marketed. All who use such equipment presumably believe that it provides the children with insight into number relationships and into the structure, enabling learning to take place. This is the essence of Gestalt theory applied to the provision of learning situations.

STRUCTURAL APPARATUS

Stern was, of course, not the first to encourage the use of structural apparatus to promote the learning of number relationships. She was, however, apparently the first to justify its use and claim its necessity on the basis of a psychological theory. There is evidence that Tillich (born 1780) and Froebel (born 1782) both advocated the use of concrete equipment in the teaching of elementary arithmetic, though this equipment may not have possessed the degree of structure which is inherent in much modern apparatus. In both cases, however, it did seem that much more than the provision of beads, counters and other unitary equipment was recommended. In particular, the Tillich equipment was concerned with a concrete approach to place-value, an idea which appears to have been taken up more recently by Dienes. Montessori (born 1870) also used a variety of forms of apparatus, including rods similar to Stern but on a much bigger scale, bead bars, counting frames, apparatus for multiplication and division, fraction equipment, and equipment for learning indices and for learning algebra.

The implications of the work of Piaget in the importance of the construction of understanding from activity and interaction with the environment, together with the growing interest in discovery learning, may have been the main reasons for the increase of interest in structural apparatus in the 1960s. Many schools procured some Stern apparatus, or Cuisenaire kits, or some of the Dienes equipment, or even some of the other kits which emerged around that time. The three mentioned specifically above are all in use today, and between them they illustrate the point that different apparatus was devised to emphasize different structures. Stern apparatus, as already mentioned, is based on segmented rods, so that eventual use of a number line for addition, for example, makes sense (Figure 6.6). Cuisenaire rods are not segmented,

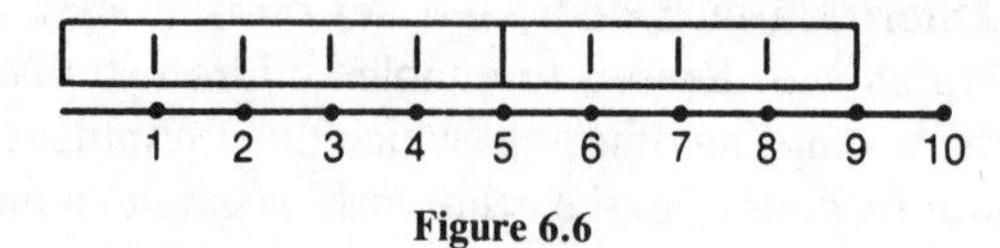

Figure 6.6

so the numerical relationships had to be learned with the aid of the colours of the rods, in relation to the lengths of the rods, rather than the number of units which they each represented. It was claimed by Cuisenaire that children grasped essential elementary number concepts better if there was no assistance from segmentation. A further advantage intended from Cuisenaire is that any rod may more easily be defined as the unit, then shorter rods become fractions. The Dienes Multi-base Arithmetic Blocks are different yet again, consisting of unit cubes, rods (similar to the segmented Stern rods), flats and blocks (larger cubes), illustrated for base five in Figure 6.7. This

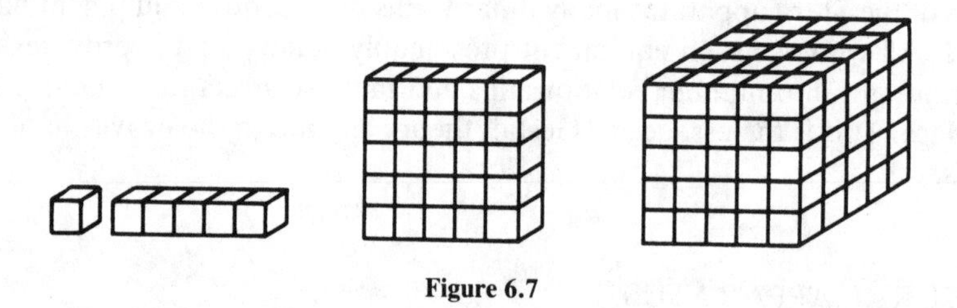

Figure 6.7

apparatus is available in all bases from two to 10 and is intended to provide insight into place-value. The Dienes Multi-base Arithmetic Blocks, of course, constitute just one type of apparatus suggested by Dienes, about whom more will be said in Chapter 9.

There might be some temptation to say that structural apparatus was one of yesterday's 'bandwagons'. However, the case for using structural apparatus is as strong as ever, and such apparatus is certainly still being sold, so it is presumably getting some use. Many teachers will not be aware of Gestalt theory, and may not know much about the discovery learning debate, yet might instinctively feel that structural apparatus can assist learning. Much of the apparatus bought today is of cheaper and more flexible plastic, for example interlocking cubes. With such cubes the teacher needs to insert the structure, so that with one class they might be assembled as rods and used in the way Stern is used, and with another class they may be used as multibase apparatus as with the Dienes equipment already described. There are, in fact, other uses for interlocking cubes. But the gains of cheapness and flexibility of use must be weighed against the loss of purpose-built apparatus which requires no assembling, in which the underlying structure is in evidence right from the start and which is not going to collapse and destroy the very structure which was being illustrated.

The case for using structural apparatus would be overwhelming if research evidence showed that it was clearly beneficial. Unfortunately, and typically with education, it is not as simple and clear-cut as that. Certain research has suggested that there are gains, but there has usually been little evidence of long-term benefits. There does not, however, seem to be any evidence that the use of structural apparatus is in any way harmful or detrimental to learning. What many experiments have revealed very clearly is the difficulties inherent in carrying out educational research. How can one ensure that the control group were completely denied any input other than what they received from their control lessons? How can one be sure that any measured effect associated with a particular teaching method is independent of the effect of particular teachers? What sort of test should be used at the end of an experiment to ensure a fair comparison between the effects of two very different experiences?

One particular well-known problem is that teachers who wish to take part in an experiment in teaching are stimulated by the whole idea and are very keen that it should succeed. Their enthusiasm is easily conveyed to the pupils, which then enhances learning. Any measured gain is therefore likely to be the outcome of an inseparable combination of the effects of new materials and methods and the total involvement and enthusiasm of the teacher. It is also well known that the most important variable factor in trying to improve learning is the quality of the teacher. In short, the measuring instruments at our disposal are not adequate to prove convincingly to what extent structural apparatus can promote learning. It might even be that many children who learn effectively without the assistance of structural apparatus do so because their natural environment provides them with the insight to enable them to make the necessary abstract connections.

Observation of younger children, in particular, suggests to most teachers that concrete activity is essential to the introduction of number and number relationships. Not only do there appear to be benefits, but a classroom atmosphere of practical activity and talk seems preferable to exposition and written practice. Some exposition will be necessary, and some written practice may be required in due course, but only after children have had plenty of opportunity to discover the structures for themselves.

PROBLEM-SOLVING AND INVESTIGATING

The desire to help learners to become better problem-solvers is a frequently expressed aim of education, and not only of mathematical education. Gagné (1970), who classed problem-solving as the highest form of learning, defined problem-solving as 'a process by which the learner discovers a combination of previously learned rules . . . [which can be applied] . . . to achieve a solution for a novel problem situation'. Here, the word 'rule' is being used in a way similar to that in which Descartes used it (quoted in Chapter 3) as anything which has been proved or established on a previous occasion. Some people believe that solving problems is the essence of mathematics learning, even to the extent of considering that the body of knowledge, which others regard as mathematics, is merely the set of tools available for the active process of problem-solving. This process, it might be said, is a creative act of striving for a specified goal based on discovering new ways of combining existing rules (knowledge). Insight is likely to be required.

Problem-solving studies have produced an immense bibliography (see, for example, Hill, 1979). According to Lester (1977), Davis said that, 'Research into human problem-solving has a well-earned reputation for being the most chaotic of all identifiable categories of human learning'. Although this should warn us to be careful about drawing conclusions, it could be said that almost any category of studies of human learning seems to produce inconclusive results!

The description 'investigation' has appeared in the literature on learning mathematics much more frequently in recent years. The exact distinction between an investigation and a problem is not all that clear. 'Problem' has a static feel about it, but activity is involved in striving for a solution. 'Investigation' has an active feel about it. An investigation could, presumably, incorporate a problem or might lead to a problem, which then demands solution. There is some suggestion in the words that a problem has a definite end-point, whether there is a solution or not, whilst an investigation might offer much more openness. If there is a distinction, common usage does not make it clear. Both words express the idea of active participation in learning, and that is what is important.

Like other aspects of active approaches to learning discussed previously, the use of problem-solving as a deliberately intended component of a mathematics curriculum involves a radical change of teaching approach from the more traditional exposition and practice of skills. The impression conveyed to learners at all levels by expository approaches to mathematics is that the subject implies a clean, logical, tidy sequence of statements, a tightly controlled crystal-clear argument. Many learners may never appreciate that the process of establishing mathematical results, theorems and rules probably involved some very untidy activity indeed. It is this untidy activity which those who are persuaded that the process side of mathematics is important are anxious to promote. Of course, the activity might turn out to be quite tidy, but the point is that we do not mind if investigations and problems lead to untidy mathematics on paper. This does not imply that our approaches to active mathematics need be unsystematic, in fact it could be argued that we should be attempting to instil systematic approaches to problem-solving in our pupils.

An early, and famous, study of problem-solving in mathematics was by Polya (1945), in which he suggested ways of improving the teaching and learning of problem-solving, an aim subsequently taken up by Wickelgren (1974). More recent research into human problem-solving abilities has drawn attention to comparisons with the use of a computer to solve problems. Problem-solving involves the processing of information, an activity for which computers are well suited, particularly when the testing out of many possibilities is involved. A considerable proportion of current research in mathematics involves lengthy computer searches, as for example in the continuing search for larger prime numbers. Such modern research perhaps confirms that problem-solving is still an essentially untidy activity, sometimes made to appear tidy only because most of the processing carried out by the computer remains concealed.

The essence of Polya's *How to Solve It* was the elaboration and justification of a self-questioning technique to be carried out by the solver. This technique involves four stages: (1) understanding the problem, (2) devising a plan, (3) carrying out the plan and (4) looking back. The first stage must not be dismissed as trivial, for it includes such essential steps as drawing a diagram and introducing suitable notation, in addition to such considerations as whether the information provided is sufficient and whether it incorporates any redundancy. The last stage involves the final checking, but also includes extension considerations such as whether the result may be generalized and whether alternative, perhaps more economical, solutions might exist. The crucial and sometimes very difficult stages are the middle two, particularly stage (2) for which creativity, inventiveness and insight might be required. In reality, the problem-solving process might involve circularity or looping, shown in Figure 6.8.

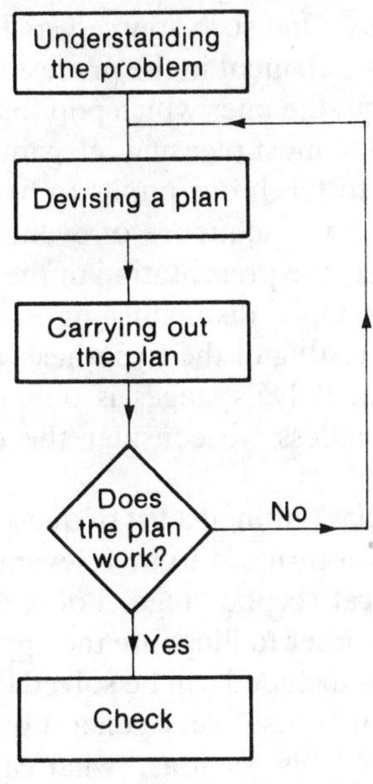

Figure 6.8

It is interesting that Hadamard (1945), drawing from writings and statements of famous mathematicians, himself included, suggested that there were four stages in the solution of a problem. His stages were: (1) preparation, (2) incubation, (3) illumination and (4) verification. The first and last of these stages are clearly similar to those described by Polya. If there is a difference it must lie in the middle two stages where Polya's implied belief, that by practising a routine, pupils and students can become better problem-solvers, might seem to be at variance with Hadamard's implication that you almost have to sit back and wait for illumination to occur. Of course, it could be said that the two were writing about different levels of problem-solving. Hadamard was writing about the process of creating new mathematics whereas Polya was writing more with the ordinary mathematics pupil or student in mind. But even restricting ourselves to the school situation one suspects there is common ground in both. For particular students and for many problems the Polya routine might well work, but there are likely to be other problems for which applications of the routine does not automatically lead to a solution. Almost everyone has experienced being unable to solve a problem at a particular moment, even after much effort but, having slept on it, or having gone away to do something different, a fruitful idea has suddenly and unexpectedly come to mind. Unfortunately, this cannot be relied on either, for there are some problems which we never solve. The incubation → illumination process is, of course, very interesting from the point of view of studies of learning. The only tenable theory so far produced is that the mind continues to search for meaning, continues to look for connections and try out connections, but at a subconscious level, thus it can produce ideas whilst we sleep and whilst we are doing other things. Sadly, this theory suggests that we might produce wonderful ideas whilst we sleep which never come to the surface when we reawaken!

There have been other, similar, studies of problem-solving. Hadamard gathered his ideas from a number of sources. One such source was Poincaré (1924) who additionally suggested a kind of aesthetic element in the process. The subconscious mind never ceases trying out connections but the ones which pop to the surface are the ones which are, in a mathematical sense, the most pleasing, elegant or even beautiful. This is not a theory which will enable us to teach our pupils to become better problem-solvers! Dewey (1910) also wrote about the sequence of events in problem-solving. He outlined five stages, which were: (1) the presentation of the problem, (2) the definition of the problem in terms of, for example, distinguishing essential features, (3) the formulation of a hypothesis, (4) the testing of the hypothesis and (5) the verification of the hypothesis. The similarity with Polya's stages is obvious. Neither Dewey's list nor Polya's list is of much value unless we consider the essential middle stages more deeply.

For each of his stages Polya has outlined a list of questions which the solver needs to ask. Not all of the questions will turn out to be relevant to all problems. The longest list was provided for the critical second stage. Polya also included many problems together with their solutions in order to illustrate the application of his procedure. Not surprisingly, all of the problems included can be solved by applying a particular subset of the procedure. There are, of course, very general questions within the procedure which are applicable in all problems, such as, 'what can we deduce from the data?' The complexity of the procedure, in that certain stages include so many questions, is likely to result in its rejection by most schoolteachers. If problem-solving can be taught we need a simpler routine, but one that does more for us than the four stage descriptions. Let us at this point consider a simple problem which does not come from Polya.

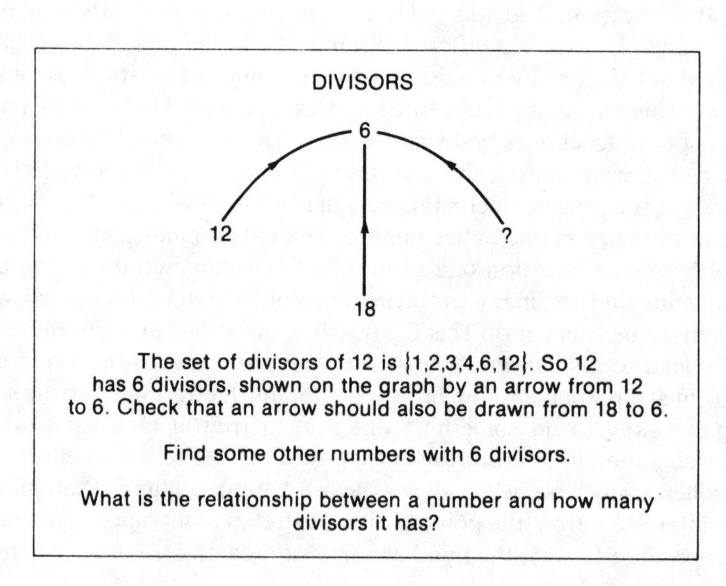

Clearly the first stage in this investigation (or in the solution of this problem) is to read and understand the statement and question, but no diagram is required and no additional notation, which dispenses with two of Polya's suggestions for stage (1). Observation of mathematics students suggests that the normal next step is to list numbers, from one upwards, together with their divisors. This might, or might not, be preceded by some rather more random trial and error attempt to find another number with six divisors, for the wording of the problem suggests that this should precede a more systematic attempt to find the relationship between numbers and how many divisors they have. The listing of numbers and their divisors provides sufficient data to enable the information to be looked at from the point of view of the divisors. Thus, for numbers 1 to 32 for example, we have the following information.

Number of divisors	Numbers
1	1
2	2, 3, 5, 7, 11, 13, 17, 19, 23, 29, 31
3	4, 9, 25
4	6, 8, 10, 14, 15, 21, 22, 26, 27
5	16
6	12, 18, 20, 28, 32
7	—
8	24, 30

We are now in a position to look for patterns in our classified results. Certain patterns stand out immediately. First, only one number (unity) has one divisor. Secondly, all the numbers with two divisors are prime. May we draw the conclusion that the only numbers with two divisors are all primes? Thirdly, all the numbers with three divisors are perfect squares. But some perfect squares apparently have more than three divisors (16, for example). What may we deduce from that information? Many numbers have four divisors, but there are at least two categories here. There are the perfect cubes (8 and 27) and there are others (6, 10, 14, 15, 21, 22, 26). What may we deduce about the others? Will all perfect cubes have exactly four divisors? And so the investigation proceeds, through a process of data collection and tabulation → pattern spotting → hypothesis formulation → hypothesis checking. This particular investigation is very relevant to the ordinary mathematics curriculum. It also illustrates a very useful procedure to apply to problem-solving, which teachers might find much more acceptable than Polya's, though we must acknowledge that Polya's work pioneered and motivated all studies of problem-solving procedures which have followed.

This illustration of problem-solving in action is similar to the illustrations contained in the recent Joint Matriculation Board/Shell Centre pack *Problems with Patterns and Numbers* (1984). If it is possible to teach children to become better problem-solvers, then the advice to teachers contained within the pack and the emphasis on a problem-solving routine will be extremely helpful. The particular list of steps in the routine promoted in this pack is:

Try some simple cases,
Find a helpful diagram,
Organize systematically,
Make a table,
Spot patterns,
Use the patterns,
Find a rule,
Check the rule,
Explain why it works.

This list applies quite well to the divisors investigation.

Generalization on the basis of patterns is an important aspect of the solution of many problems, and was fully considered by Polya. A simple example of this is in the search for a formula for the sum of cubes of natural numbers. Trying out simple cases systematically produces

$$\begin{aligned} 1^3 &= 1 \\ 1^3 + 2^3 &= 9 \\ 1^3 + 2^3 + 3^3 &= 36 \\ 1^3 + 2^3 + 3^3 + 4^3 &= 100 \\ 1^3 + 2^3 + 3^3 + 4^3 + 5^3 &= 225. \end{aligned}$$

The totals should immediately suggest squares, for $1 = 1^2$, $9 = 3^2$, $36 = 6^2$, $100 = 10^2$ and $225 = 15^2$. The numbers 1, 3, 6, 10 and 15, the first five triangular numbers, are the sums of consecutive natural numbers,

$$\begin{aligned} 1 &= 1 \\ 1 + 2 &= 3 \\ 1 + 2 + 3 &= 6 \\ 1 + 2 + 3 + 4 &= 10 \\ 1 + 2 + 3 + 4 + 5 &= 15, \end{aligned}$$

which then suggests the generalization

$$1^3 + 2^3 + \ldots\ldots\ldots\ldots + n^3 = (1 + 2 + \ldots\ldots + n)^2.$$

To complete the process a mathematician would now require a proof of the generalization. The usual method of proof for this result is induction, a method which many students, meeting it for the first time, regard with great suspicion. There is, for them, a feeling of having assumed the result which they were expected to prove and that, having made the assumption, the proof is not a proof. This is an interesting example of a barrier occurring in the mind. There are better-known barriers to problem-solving which are considered in the next section of this chapter.

Polya's *How to Solve It* included a discussion of all of the standard methods of proof in mathematics, that is, induction, deduction, contradiction, counter example and working backwards. He also elaborated a number of other aspects of problem-solving to which his outline procedure made reference, like reducing the dimensions of the problem, looking for methods of solving similar problems, and relaxing some of the conditions associated with the problem. Polya continued his work in two further books (1954 and 1962). Wickelgren (1974) based his work on the same premise as Polya, namely that problem-solving in mathematics *can* be taught, and indeed is careful to acknowledge the pioneering work of Polya. Wickelgren maintained that there were only seven types of mathematical problem in the sense that, in order to achieve a solution, it was necessary to use one or other of the seven methods which he explained in considerable detail. Each of these seven methods, he claimed, needed to be practised, if one wished to become a better problem-solver. Wickelgren also included an enormous number of problems in his book, which are of great interest to teachers of mathematics.

In recent years many books have been published which are concerned with helping teachers to take problem-solving into the classroom (see for example Burton, 1984). It might well be that pupils can be taught to become better problem-solvers but, in carrying out teaching, it is helpful if we are aware of the well-known difficulties which might arise.

OBSTACLES AND DIFFICULTIES IN PROBLEM-SOLVING

There are a number of well-known obstacles and difficulties which might hinder attempts to solve problems. These will be discussed with reference to the following six problems which the reader should attempt before reading on.

Problem 1

• • •

• • •

• • •

Using only four straight lines, connect the nine dots without lifting pen from paper.

Problem 2

Assemble six matches so that they form four congruent equilateral triangles each side (edge) of which is equal to the length of the matches.

| | | | | |

Problem 3

Four soldiers have to cross a river. The only means of transportation is a small boat in which two boys are playing. The boat can carry at most two boys or one soldier. How can the soldiers cross to the other side?

Problem 4

Given 3 containers ('buckets') and plenty of water, the task is to measure out a required amount.
Example: Given 3,21 and 127 litre buckets, measure out 100 litres.

Solution: 100 = 127 − 21 − 3 − 3.

Problems:

	Buckets			*Goal*
	a	*b*	*c*	
1	21	127	3	100
2	14	46	5	22
3	18	43	10	5
4	7	42	6	23
5	20	57	4	29
6	23	49	3	20
7	15	39	3	18

Problem 5

Given the sum

```
  D O N A L D
+ G E R A L D
-------------
  R O B E R T
```

and the fact that D = 5, find what numbers the letters represent. Every digit 0 to 9 has a different letter, every letter has a different number.

The problems above, or variations of them, are so widely used that it is difficult to know where they originated. They have been used, however, by a variety of researchers involved in investigating human problem-solving behaviour.

Very few people, including university students of mathematics, are able to solve Problem 1 within a limited time period. It is necessary to draw lines which go outside the square implicitly defined by the nine dots (see Figure 6.9). It appears that most people are not prepared to do that; they appear to work under the assumption that the four lines must lie within the square. No such restriction was stated in the question. Why this phenomenon should occur is perhaps not fully understood, but it does happen in problem-solving. The phenomenon was discussed by Scheerer (1963) as an

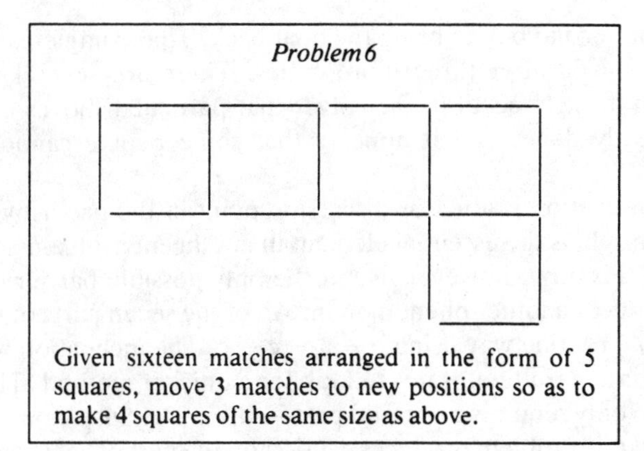

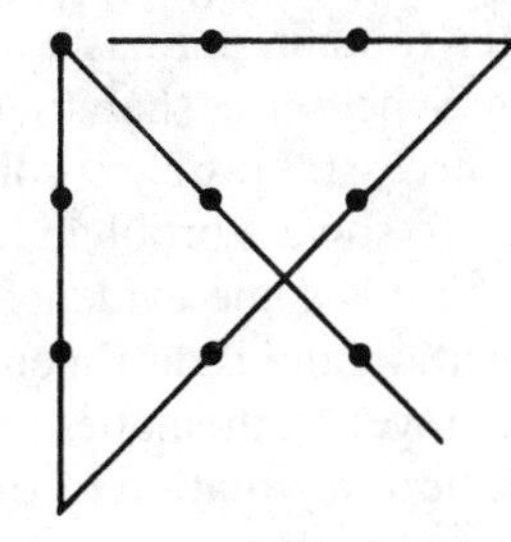

Figure 6.9

example of 'fixation', sometimes referred to as 'problem-solving set'. It appears that we are prone to making initial assumptions which are not included in the problem specification.

The second problem also brings to light another fixation. Most people try to arrange the matches on a flat surface and fail to appreciate the novel three-dimensional use which is required. The solution (shown in Figure 6.10) is to arrange the matches into a tetrahedron.

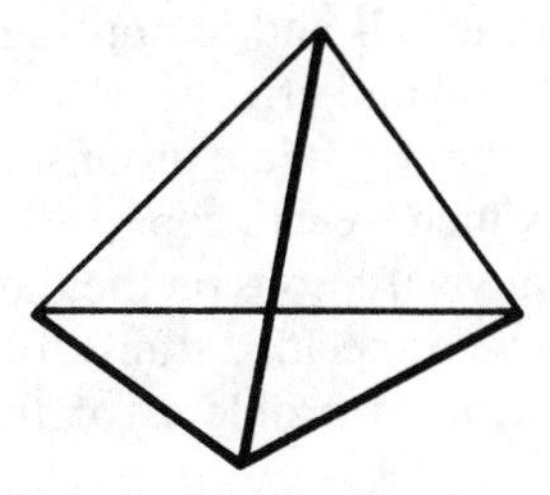

Figure 6.10

The third problem, noticeably rather less of a problem than the other two as far as mathematics students are concerned, depends on an iteration or cycle of moves. The fixation here is described by Scheerer as unwillingness to accept a detour when it appears that the achievement of the required goal is being delayed. The solution requires the two boys to row across, one to return with the boat, one soldier to cross in the boat, and the second boy to bring the boat back. That completes the first cycle, so the problem is solved after three more cycles. There are several moments in this procedure when it might seem to the solver that particular moves undo some of the good work already done, and it appears that some people cannot surmount that barrier.

The barrier of fixation is a major discussion point in the paper by Scheerer (1963) who wrote: 'If insight is the essential element in intelligent problem-solving, fixation is its arch enemy.' Fixation, however, is not the only possible barrier or difficulty, and Problem 4 introduces another phenomenon. All of the seven parts of the problem may be completed in a routine way using $b - a - c - c$. The inclination we all have, once we have settled into a routine, is not to look for a quicker method. Thus we might fail to see that part 6 only requires $a - c$ and part 7 only $a + c$. Suspicion that there is some kind of 'catch' in this problem produces caution on the part of some solvers which then enables them to see the quicker solutions. In general, in mathematics, we are sometimes inclined to overlook a quicker route in our haste to apply a known routine. This particular barrier was described by Scheerer as 'habituation'.

There is yet another barrier to successful problem-solving, known as 'over-motivation'. Sometimes we are motivated to solve a problem almost to the point of anxiety. Scheerer's explanation was that 'there is some evidence that strong ego-involvement in a problem makes for over-motivation and is detrimental to a solution'. Thus, when a pupil presents the teacher with a novel mathematics problem and then waits around to watch over the attempts at solution, a situation of over-motivation can arise! But this barrier, and the others mentioned earlier, must affect pupils too.

Problem 5 was discussed by Bartlett (1958) in the context of a classification of problem types. Newell and Simon (1972), however, looked at how subjects attempted to solve the problem. They found that subjects who achieved the solution all followed very similar paths, based on processing the most constrained columns in order of emergence. Thus $T = 0$, $E = 9$, $A = 4$, $R = 7$, $G = 1$, $L = 8$, $N = 6$, $B = 3$ and $O = 2$. In this problem there is little need for much more than serial processing which places little demand on short-term memory. Thus, we cope with this problem quite well. If, however, we tried to solve the problem by investigating the number of possible assignments of digits to letters we would find the demands of the search procedure unacceptable. In general terms, Newell and Simon suggested that we are willing to endure only a limited amount of trial and error search. So, in situations which involve a huge 'problem-space' it is necessary to isolate promising parts of the space, or to first find promising approaches with which we can cope.

The sixth problem is one of many variations on the theme of moving matches, sticks or other objects in order to achieve a specified transformation. The expected solution is shown in Figure 6.11. Katona used problems of this type in Gestalt studies of

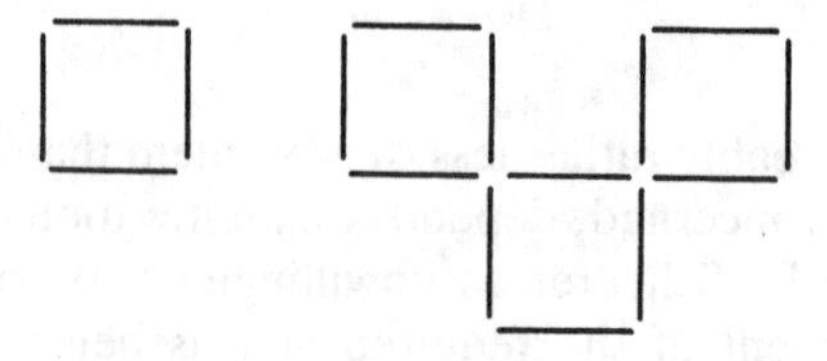

Figure 6.11

problem-solving (Katona, 1940). Of three methods used to try to promote successful problem-solving he discovered that the least effective was to demonstrate the solution and then rely on the solver remembering the procedure. Such a method is dependent on an isolated unit of knowledge which cannot be related to units of existing knowledge structured in the mind. It is an example of rote learning and it should be no surprise that it was comparatively unsuccessful. There are surely lessons to be learned here in terms of the value of showing a pupil how to solve a mathematics problem. A second method used by Katona was to make helpful statements like, for example, 'matches with double functions should be moved so that they have single functions', and 'proceed by creating holes and generally loosening the figures'. This method was reasonably effective in promoting success but the best teaching method was to illustrate what changes would be brought about by particular moves of individual matches, such as moving one with a double function. It is interesting that both the second and third methods were based on providing limited guidance. The second was based on making verbal statements but the third, the most successful, was based on carrying out particular actions and forcing the student to try out particular moves and think about their effects.

Newell and Simon (1972) have also pointed out difficulties in problem-solving which stem from the limitations of the human information-processing system. First, we can only operate in a serial manner, one process at a time, and not on several processes in parallel. Secondly, whatever processing we do has to pass through our short-term memory, which has a limited capacity of around seven units of information (Miller, 1956). Sometimes it might be possible to retain more in our short-term memory by chunking units of knowledge, but we are still limited. Thirdly, as regards long-term memory, although we have apparently virtually unlimited storage capacity, there is the difficulty of committing knowledge to long-term memory. It takes a long time to do this in comparison with the very short time required for actual processing. It can be difficult to do it unless we can link the new knowledge to existing knowledge. Retrieval of knowledge held in long-term memory is by no means automatic. And then there is the phenomenon of forgetting (see Chapter 9).

LOGO

In elementary geometrical studies children are introduced to polygons. The task of constructing a regular pentagon requires knowledge about angles, either angles at the centre, or interior angles or exterior angles. If we wish children to discover the angle facts which are required to construct a regular pentagon we would be in some difficulty. Trial and error might prove tedious, and computation of angles is likely to be considerably beyond the capabilities of young children. By using a computer equipped with LOGO facilities, however, it is claimed that children are able to discover how to draw such shapes as regular pentagons. Trial and error is involved but, possibly because they are using a computer which does all the tedious work, children do not lose their motivation. LOGO allows children to construct their own knowledge, in this case about polygons, through a process of discovery.

LOGO is a computer language, though certain proponents might prefer the description 'learning environment'. It originated in the USA around 1967, but schools have had to wait until the relatively recent introduction of microcomputers to make use of what it offers. It is, currently, perhaps the most obvious example of the use of a computer to facilitate the discovery of mathematical knowledge. The instigator of school activities using LOGO was Papert, who also pioneered the use of a mobile 'turtle' which can be used to trace out shapes on a classroom floor and which, in so doing, captivates most children. Papert gained his enthusiasm for active, discovery-type learning directly from Piaget, with whom he worked for five years. Summing up his belief he said he left, 'impressed by [Piaget's] way of looking at children as the active builders of their own intellectual structures' (Papert, 1980), but he was inclined to think that the enrichment of the learning environment through the use of materials was of greater significance than Piaget had suggested. The power of LOGO as a learning environment was thus harnessed by Papert to facilitate children's construction of their own knowledge.

The basic geometrical use of LOGO allows children to create their own shapes and, by a process of successive refinement, produce particular pre-determined shapes. A 'turtle' is not essential; the shapes may be drawn on the screen of a television set. Basic commands such as **forward** and **backward** produce lines, and turns are effected by typing in commands like **left** and **right**. Thus:

```
forward  40
right    90
forward  40
right    90
forward  40
right    90
forward  40
right    90
```

produces a square and leaves the moving point ready to set off in its original direction. Such a program may be abbreviated, basically by instructing the computer to repeat a pair of instructions four times; it may also be generalized both in terms of the length of the sides and also in terms of the finished shape, by altering the angle of turn. To watch children trying to produce an equilateral triangle by using first one angle of turn, and then another, and so on, until the correct one is discovered through a combination of trial and error and genuine thinking is an experience in watching discovery taking place. In trying to draw the simple house in Figure 6.12 an enormous

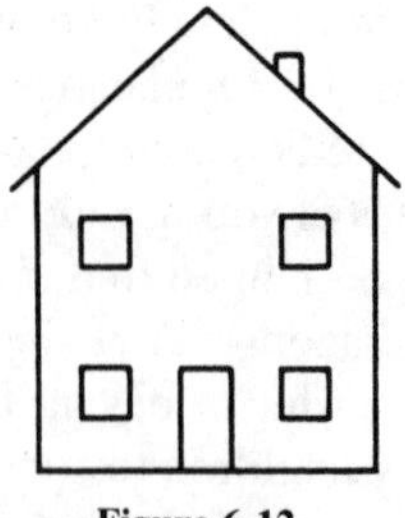

Figure 6.12

amount can be learned about simple geometry. This is only the beginning of what

LOGO has to offer in learning geometry, which in itself, we are given to understand, is only the beginning of what LOGO has to offer in the promotion of learning. Most of the research which has been carried out into the value of LOGO as a learning environment has been carried out with young children. The value of LOGO both with older children and outside geometry remains comparatively unexplored, though developments are taking place.

Papert's approach to the application of Piagetian theory to education involved a mildly critical attack on the perceived implications of the acceptance of belief in concrete and formal thinking. Papert claimed that the boundary between the two could be moved, with the help of the computer, and that 'the computer can concretize (and personalize) the formal'. The computer provided the additional advantage that the child's anxiety level is reduced, and the desire to obtain the correct answer is not so compelling. Working with LOGO not only guaranteed many wrong answers, producing merely impersonal and uncritical responses from the computer, it also encouraged the process of 'debugging', claimed by Papert as an important life skill. Papert believed that the usual mathematics curriculum was meaningless to most children, but LOGO allowed them to construct knowledge in a meaningful way. He even linked his beliefs with the work of Polya in problem-solving in saying, 'the best way to explain Polya to students is to let them learn Turtle Geometry'.

A considerable amount of research into the use of LOGO with children has now been carried out. In Britain most of this research has not started from the extreme premise of Papert, but has been more concerned with integrating LOGO into the existing curriculum, though Papert considered such work as a misrepresentation of what he was attempting to advocate. In one particular study, carried out at Edinburgh University, the most outstanding benefits which were discovered were, in fact, not that the normal curriculum was mastered more thoroughly with the help of LOGO, but that the pupils had become better able to discuss and argue about mathematics, and that their problem-solving capabilities had been enhanced. In the USA research has been much more commonly concerned with experimentation based on the claims of Papert. For example, the Lamplighter project in Dallas attempted to provide an environment for the children which was, given the constraints of modern schooling, as close as possible to that envisaged by Papert.

We do not yet have a body of evidence which authentically vouches for the value of LOGO as a learning environment. Almost certainly it will not be used on a large scale in the way that Papert might have wished, but it is undoubtedly an excellent example of the way the computer might be used to allow children to discover mathematics for themselves.

SUGGESTIONS FOR FURTHER READING

Ausubel, D. P. (1963) Some psychological and educational limitations of learning by discovery. *New York State Mathematics Teachers Journal* **XIII**, 90–108. (Also in *The Arithmetic Teacher* **11** (1964), 290–302.)

Biggs, E. E. (1972) Investigational methods. In L. R. Chapman (ed.), *The Process of Learning Mathematics*. Oxford: Pergamon Press.

Bruner, J. S. (1960) On learning mathematics, *The Mathematics Teacher* **53**, 610–19.

Papert, S. (1980) *Mindstorms*. Brighton: Harvester Press.

Polya, G. (1945) *How to Solve It*. Princeton: Princeton University Press.
Scheerer, M. (1963) Problem-solving, *Scientific American* **208** (4), 118–28.
Wickelgren, W. (1974) *How to Solve Problems*. San Francisco: Freeman.

QUESTIONS FOR DISCUSSION

1 Under what conditions are children able to discover mathematics for themselves?
2 Is problem-solving the essence of mathematics?
3 What is the place of structural apparatus throughout the period of compulsory schooling?
4 How might the computer be used to promote active rather than passive learning?

Chapter 7

Why Do Some Pupils Achieve More Than Others?

INDIVIDUAL DIFFERENCES

A wide variety of differences between pupils will be observed when we ask them to do mathematics. How, for example, are pupils likely to react to this problem?

> *League fixtures*
>
> There are eight teams in a Junior League. How many League matches are required in completing all the fixtures in one season? If there were n teams, how many matches would be required?

Some children play in a local Junior League in some sport or other, so the context would be familiar to them. They might even have been so motivated to have obtained the solution independent of mathematics lessons and long before the teacher posed the question. They, and some children who do not play in a local league, would find the problem interesting and worth attempting, but many other pupils would not be at all interested. It might be found that more boys than girls are motivated by the problem, but it might not. Some contexts are more immediately appealing to girls and others to boys. Some children, despite finding the context interesting, might not be able to solve the problem. Amongst those who can solve the problem, a variety of different methods might be used. The fact that so much variety will undoubtedly emerge in response to one particular mathematical situation is another feature which needs to be taken into account in considering the learning process.

For those pupils who are unable to solve the problem the teacher might decide to demonstrate or discuss a solution. There are, however, so many different ways of setting about the problem that it is difficult to know which is the best method to use. Whichever method is chosen, it might not be the best method for many pupils. A variety of methods of solution is outlined below, all based on numbering the teams one to eight.

Method A

The fixtures for Team 1 are:

1 *v*. 2   2 *v*. 1
1 *v*. 3   3 *v*. 1
1 *v*. 4   4 *v*. 1
1 *v*. 5   5 *v*. 1
1 *v*. 6   6 *v*. 1
1 *v*. 7   7 *v*. 1
1 *v*. 8   8 *v*. 1

so the total number of matches involving Team 1 is 14.

The fixtures for Team 2 are:

2 *v*. 1   1 *v*. 2
2 *v*. 3   3 *v*. 2
2 *v*. 4   4 *v*. 2
2 *v*. 5   5 *v*. 2
2 *v*. 6   6 *v*. 2
2 *v*. 7   7 *v*. 2
2 *v*. 8   8 *v*. 2

but two of these fixtures, 2 *v*. 1 and 1 *v*. 2, have already been counted amongst the fixtures for Team 1, so there are only 12 new matches.

Proceeding by listing might eventually lead to shortcutting, though it might not, but eventually the method results in a total number of matches of:

$$14 + 12 + 10 + 8 + 6 + 4 + 2 = 56$$

Method B

Each team plays all of the other seven teams twice, thus each team plays 14 matches. This method counts each match twice, since 1 *v*. 2, for example, is counted both as a home match for Team 1 and an away match for Team 2. The total number of matches is therefore:

$$14 \times 8 \times \tfrac{1}{2}$$
$$= 56$$

Method C

All we need to do is find how many home matches each team plays and total them. Each team plays seven home matches, so the total number of matches is:

$$7 \times 8$$
$$= 56$$

Mathematically, this might be considered the same as Method B, but pupils might not appreciate that.

Method D

There are eight numbers, 1, 2, 3, 4, 5, 6, 7 and 8, and we need to know how many pairs of numbers may be selected from these eight. This total is, in fact, $8 \times 8 = 64$, but this includes 1 *v.* 1, 2 *v.* 2, etc. The number of matches is therefore:

$$8 \times 8 - 8$$
$$= 56$$

When the number of teams is allowed to vary, the generalization also permits variation of method.

Method X

We know that a league of eight teams produces 56 matches. Nine teams, that is, introducing a ninth team on to the already existing eight, introduces another eight home matches and another eight away matches giving a total number of fixtures of:

$$56 + 2 \times 8$$
$$= 72$$

Then 10 teams produce a total of:

$$72 + 2 \times 9$$
$$= 90$$

and so on.

Method Y

Any of methods A to D applied to a different number of teams, to provide the same data as in Method X.

Method Z

A tabulation of all numbers of fixtures for any number of teams, starting with one team, so that a number pattern emerges, which may be extended using, for example, the fact that the differences are consecutive even numbers.

Number of teams	*Number of fixtures*
1	0
2	2
3	6
4	12
5	20
6	30
7	42
8	56
9	72
10	90

From the tabulated data, however that data was obtained, a generalization for n teams is required. It might be clear to some pupils that the number of fixtures is always:

(number of teams) × (one less than the number of teams).

Other pupils might see things differently, in that the number of fixtures is:

(the square of the number of teams) − (the number of teams).

Yet more pupils, who have successfully produced results like those in the table, will not be able to proceed with the use of letters, and will find $n(n-1)$ unacceptable.

This issue has been explored at length because the fact that different pupils will think their way through mathematical problems in a variety of different ways is a complication presented to all teachers. For two reasons any taught method for solving a particular problem might not meet the requirements of all pupils. First, it might not meet all of them where they are (Ausubel, 1968), in the sense that too many assumptions are made about prior knowledge. Secondly, it might not coincide with their preferred cognitive style. Discussion of a variety of different methods of solving the same problem might be thought to be much more beneficial. On the other hand such an approach might be rejected on a number of grounds including boredom created by spending so long on the one problem and lack of interest amongst those pupils who have decided that their method of solution is the only one that matters.

The issue of different methods favoured by different pupils on the same problem is only one within the complete range of issues associated with individual differences. Some pupils clearly do achieve more in their studies of mathematics than do others, hence there *are* differences. Abilities, preferences, attitudes and motivation all contribute to making some pupils more successful than others, and in the remainder of this chapter a variety of contributory factors are considered.

CONVERGENT AND DIVERGENT THINKING

It is necessary to consider the use of convergent and divergent thinking at this stage in order that reference can be made to the ideas in subsequent sections. The convergence/divergence dichotomy (or is it a spectrum?) is best introduced by means of examples.

Tests of intelligence are of many types but some, and particularly tests of non-verbal intelligence, include questions which are numerical, for example:

> 1, 5, 9, 13, 17 . . .
>
> What number comes next?

and

> 1, 2, 3, 4, 5, 6, 7, 8, 9.
>
> Write down the difference between the largest and the smallest of these numbers.

also

> Here are three figures:
>
> 5, 2, 7.
>
> Add the largest two figures together and divide the total by the smallest.

Some intelligence tests also, or alternatively, involve diagrams, as in these further examples:

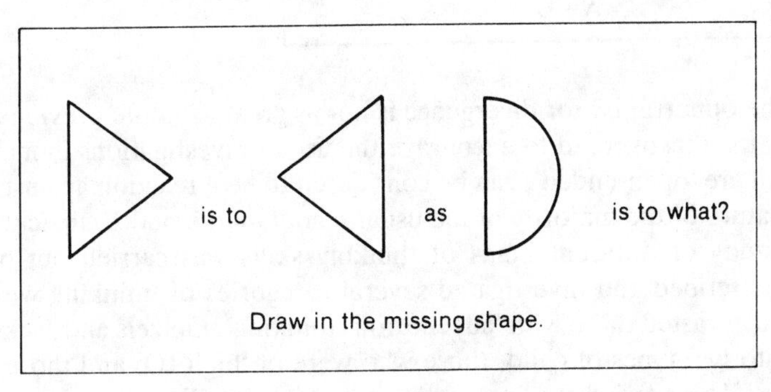

and

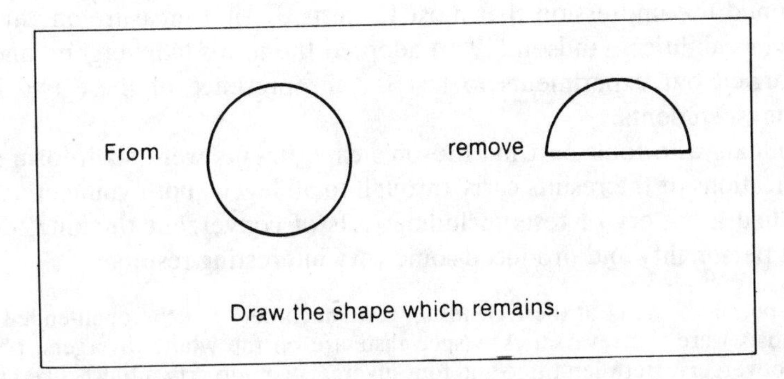

A good example of a test which includes items of the numerical and diagrammatic kinds represented here is that by Heim (1970).

All of these sample test items might be said to be mathematical. In addition, for all of the items, there is only one expected answer, that is, the questions are all convergent. To most people, mathematics appears to be a subject in which convergent thinking skills are the most valued. In fact, there may be no evidence that divergent thinking skills are needed at all. In other curriculum areas it is very easy to produce questions which are divergent, that is, which provide the opportunity for many varieties of acceptable responses. The following test item certainly allows divergence:

> Write down as many words as you can which end in
>
> -ing.

Test items in mathematics which are divergent are not much used. One which Guilford (1959) included as an example was:

> Make up as many equations as you can which follow from
>
> $$B - C = D$$
>
> and
>
> $$Z = A + D$$

Even then, the opportunity for divergence is not as great as in the previous test item. In one sense, the recent trend to encourage the use of investigations in mathematics, many of which are 'open-ended', can be considered to be a reaction against the highly convergent nature of the majority of the usual school mathematics curriculum.

An early study of different kinds of thinking skills was carried out by Bartlett (1958), who described and investigated several categories of 'thinking within closed systems' and contrasted these with 'adventurous thinking'. Getzels and Jackson (1962) researched into two types of child, those who were of 'high IQ' and those who were 'high creative'. This carried the implication that high intelligence only depended on convergent thinking skills and that creativity had nothing to do with intelligence. In fact it confirmed the impression that most IQ tests did not measure anything other than convergent abilities. Hudson (1966) adopted the terms 'convergent' and 'divergent' and carried out experiments to test for the presence of these two kinds of abilities amongst students.

The individuals who took part in Hudson's experiments were sixth-form students but the implications of the results carry through to all levels, both younger and older. Hudson applied a battery of tests including tests of convergent thinking, divergent thinking and personality and produced some very interesting results:

> Most arts specialists, weak at the I.Q. tests, were much better at the open-ended ones; most scientists were the reverse. Arts specialists are on the whole divergers, physical scientists convergers. Between three and four divergers go into arts subjects like history, English literature and modern languages for every one that goes into physical science. And, vice-versa, between three and four convergers do mathematics, physics and chemistry for every one that goes into the arts.

Certain traditional school subjects did not fall into just one category, for example biology, geography and economics attracted convergers and divergers roughly equally. Classics, on the other hand, went with the physical sciences and mathematics.

Only a minority of students coped equally well with convergent and divergent items. Such results demand further reflection.

If most mathematics students are predominantly convergent thinkers, does this imply that few specialist mathematicians are capable of creativity or inventiveness? It cannot be the case that no mathematics students are genuinely creative, but Hudson's results might be taken to suggest that the only creative mathematics students are the minority who are either divergent thinkers or who are equally at home with both kinds of thinking! Further, does the predominantly convergent nature of the thinking of the majority of mathematics sixth-form students imply that this will remain the case throughout life, or is there hope for us all? Does the typical school mathematics education produce the convergent thinkers which we find in the sixth form or is it the case that the pupils are already predisposed towards convergence and our mathematics curriculum does little to counteract it? Are students attracted to mathematics because it appears that only convergent thinking is required? If we attempt radical changes to the mathematics curriculum so that it becomes much more based on open-ended situations will we deter some students whose preference is for convergent studies? What should we be doing, as mathematics teachers, to cater for both convergence and divergence in the preferences of our pupils?

In this particular domain of individual differences there may seem to be more questions than answers at the moment. Hudson pursued his research into the affective domain by looking for correlations between the convergence/divergence trait and personality traits. As a result of this research he suggested that convergent thinking is the preference of the pupil who likes to keep emotion apart from studies and that divergent thinking is the preference of those who like their studies to involve them emotionally. In short, the theory proposed by Hudson was that affective predispositions which reach right back into young childhood cause us to prefer either convergent- or divergent-related studies, so these are the kinds of studies in which we subsequently choose to specialize. It may be that implications can be tested by looking, for example, at mathematics teachers as compared with teachers of certain other subjects! Of course, some of those who choose to specialize in mathematics are at ease with both convergent and divergent situations. The theory cannot, however, be dismissed lightly. The following statements from two 17-year-old pupils were collected by the author and by Russell (1983).

> I think the popularity of maths depends very much on the character of the person. Maths is an ideally suited subject to anyone who likes logic, clear-cut solutions, methods, definite right or wrong answers. Such people are bound to enjoy mathematics.

> Maths doesn't show your personality. English is very much you—far more of your personality comes through.

MATHEMATICAL ABILITY

The analysis of human abilities has been the subject of many studies which have taken a variety of different forms. At one extreme has been the method based on the statistical procedure of factor analysis applied to test scores. At the other extreme is the anecdotal approach, often based on the reflections by famous mathematicians and others about their own ability. The outcome has been the clear indication that overall intellectual capacity is the most dominant influence on mathematical ability, and it is a matter of what other, more specific, abilities can be shown to exist. Some pupils, however, clearly do show more aptitude for mathematics than others, so the issue of mathematical ability is essential to a consideration of individual differences. Particular points of interest, apart from what it is that makes one pupil more able than another, include whether it is possible to identify pupils of high mathematical ability early in life and whether it is possible to foster such ability.

A major study of mathematical ability in pupils was carried out by Krutetskii (1976). The study was, in essence, based on observation of, and conversation with, pupils—so it is not surprising that the research method used by Krutetskii has been compared with that of Piaget. The origins of mathematical ability were seen by Krutetskii to lie in the existence of 'inborn inclinations', as can be seen in the following statements:

> Mathematical abilities are not innate, but are properties acquired in life that are formed on the basis of certain inclinations . . . some persons have inborn characteristics in the structure and functional features of their brains which are extremely favourable to the development of mathematical abilities . . . anyone can become an ordinary mathematician; one must be born an outstandingly talented one.

This theory must be seen against the Soviet political background which demanded that individual differences should be explained without allowing mathematical abilities to be innate. The resulting confusion between the relative contributions from innateness and from environmental factors, contained in Krutetskii's book, does not provide a secure foundation for explaining mathematical ability. In other ways, however, Krutetskii's work is helpful.

One definition of mathematical ability stated by Krutetskii was:

> individual psychological characteristics . . . that answer the requirements of school mathematical activity and that influence . . . success in the creative mastery of mathematics as a school subject—in particular, a relatively rapid, easy, and thorough mastery of knowledge, skills, and habits in mathematics.

In more detail, the components of mathematical ability were seen by Krutetskii to be:

1. An ability to extract the formal structure from the content of a mathematical problem and to operate with that formal structure,
2. An ability to generalize from mathematical results,
3. An ability to operate with symbols, including numbers,
4. An ability for spatial concepts, required in certain branches of mathematics,
5. A logical reasoning ability,
6. An ability to shorten the process of reasoning,
7. An ability to be flexible in switching from one approach to another, including both the avoidance of fixations [see chapter 6 of this book] and the ability to reverse trains of thought,
8. An ability to achieve clarity, simplicity, economy and rationality in mathematical argument and proof,
9. A good memory for mathematical knowledge and ideas.

It is relevant to compare this analysis with that described by Suydam and Weaver (1977), reflecting on characteristics of good problem-solvers in mathematics:

1. Ability to estimate and analyze,
2. Ability to visualize and interpret quantitative facts and relationships,
3. Ability to understand mathematical terms and concepts,
4. Ability to note likenesses, differences and analogies,
5. Ability to select correct procedures and data,
6. Ability to note irrelevant detail,
7. Ability to generalize on the basis of few examples,
8. Ability to switch methods readily,
9. Higher scores for self-esteem and lower scores for text anxiety.

Suydam and Weaver also noted that 'more impulsive students are often poor problem-solvers, while more reflective students are likely to be good problem-solvers'.

It might be considered reasonable to assume that the existence or otherwise of mathematical ability stems from physiological sources. Krutetskii referred to physiological considerations on several occasions. Mathematically able pupils adopt a procedure in solving mathematical problems which suggests they can follow a plan which involves trying out ideas systematically and in which they appear to be able to see which ideas are worth pursuing and which are not. Incapable pupils, on the other hand, show 'blind, unmotivated manipulations, chaotic and unsystematic attempts to find a solution'. The physiological explanation provided by Krutetskii was that, within the cortex of the brain, there is a control apparatus, the 'acceptor of an operation', which evaluates results of any operation, comparing what has been tried with what could be tried and generally directing and regulating trials. Thus, 'when there is a pronounced inability for mathematics, a low level of functional maturity of the inferior parietal region of the cortex and of its connections with other sections of the brain is observed'.

Krutetskii also suggested that there were different kinds of mathematical ability. Some pupils had an 'analytic' mind and preferred to think in verbal, logical ways. Other pupils had a 'geometric' mind and liked a visual or pictorial approach. And yet other pupils had a 'harmonic' mind and were able to combine characteristics of both the analytic and the geometric, though they were likely to show some leaning towards either the analytic or the geometric approach. Pupils with a harmonic type of mind were most likely to show real mathematical aptitude. The suggestion that there were varieties of 'mathematical mind' had been made earlier by, for example, Hadamard (1945).

Earlier attempts to describe mathematical ability made by Hamley, Haecker and Ziehen, Oldham and Werdelin were reviewed by Krutetskii. Hamley had concluded that 'mathematical ability is probably a compound of general intelligence, visual imagery, ability to perceive number and space configurations, and to retain such considerations'. Haecker and Ziehen had provided a schedule of components of mathematical ability which were derived from four main components, namely spatial, logical, numerical and symbolic. Oldham had concluded that the concept of one single entity which might be described as mathematical ability was an elusive one. Werdelin had concluded that it was the ability for reasoning in mathematics which was the key to mathematical ability. None of these studies has provided particularly valuable information about mathematical ability.

Hadamard (1945), in using evidence from studies of a number of famous mathematicians of his day and of earlier times, considered it doubtful that mathematical aptitude existed as separate from aptitude generally. He justified this view in two ways. First, he pointed out that few pupils who excelled in mathematics at school were useless in other areas of human knowledge. Secondly, he stated that many creative mathematicians had also been creative in other spheres of study, and he quoted Gauss, Newton, Descartes and Leibniz in support of this view. It must be admitted, however, that many mathematicians who contributed in other fields did so in closely allied subject areas like physical science. Others took their mathematical reasoning powers into logic and philosophy. Nevertheless, Hadamard might be correct in general terms, and there certainly are examples of eminent contributors to very varied fields of study, like Leonardo da Vinci and Lewis Carroll. The widely quoted correlation between mathematics and music is also of interest in this context. Drawing conclusions from a few examples or from anecdotal evidence is always dangerous, but few teachers would disagree that the most mathematically talented pupils usually have many other talents too.

Just as Krutetskii pointed out differences between mathematically able pupils, so too Hadamard cited differences between famous mathematicians. Riemann was said to have had an 'intuitive' mind, whilst Weierstrass was 'logical'. Hermite preferred analysis whilst Hadamard himself thought 'geometrically'. Poincaré is said to have claimed that he could not carry out an addition without making a mistake! Hadamard claimed to have experienced great difficulty in mastering certain mathematical ideas, like group theory, whilst being able to contribute original ideas in other branches of mathematics. Einstein, of course, has been described as having been useless at anything other than mathematics and physics, with even a suspicion that his mathematical talents were limited, demonstrating not only a restricted facility within mathematics but also that not all of the mathematically talented have talents outside mathematics.

The other phenomenon which needs to be taken into account is the 'prodigious calculator'. Hadamard suggested that this is a separate and distinct ability from mathematical ability because many prodigious calculators did not appear to be in any real sense mathematicians. Yet, in identifying pupils who are highly talented in mathematics, one of the signs to look for is said to be a fascination with numbers and high facility in handling them. The issue of the prodigious calculator has been considered more recently by Hope (1985). Many children become almost obsessed with one particular interest, such as sport, and are so motivated as to make themselves experts in that interest, with wide knowledge of the facts and statistics. In the same way, Hope suggested, some children's early fascination with number and number relationships motivates them to learn and memorize much more than is normal even for an able child, eventually resulting in the prodigious calculator. The characteristics of such a person, as described by Hope, are mastery of many number facts committed to long-term memory, excellent short-term memory with a capacity greater than the normal 7 ± 2 units (Miller, 1956), and mastery of methods of processing which make the best use of our limited short-term memory, for example left to right calculation with replacement of running totals instead of the taught method of right to left calculation. There is also the likelihood that the prodigious calculator will be making use of mathematical relationships which many pupils meet but of which few realize the value in calculations, for example $a^2 - b^2 = (a - b)(a + b)$ which enables computations such as $63^2 - 37^2$ to be completed easily. Such expert calculators are frequently in the news, usually in entertainment, but, of the many which history has recorded, only Gauss and Aitken have been considered to be mathematicians.

Hadamard's work was concerned with the mathematical ability of great mathematicians. Throughout this century, psychologists have attempted to investigate ability, both overall intelligence and also specific abilities. Many such psychological studies have been based on the statistical techniques of factor analysis (see for example Vernon, 1950), though the basis of factor analysis has always been disputed. Generally, such studies have not produced evidence to contradict the well-known theory of Spearman that a general intellectual factor (denoted by 'g') operates across all domains of human intellectual activity. In fact, factor analysis often suggests that the 'g' factor is dominant. Whether such separate abilities as mathematical ability, geographical ability, historical ability and the like exist has not been fully substantiated, yet we do talk about people as having great musical or artistic ability, and even great mathematical ability. Factor-analytic studies have been used to justify the existence of group factors, such as verbal ability, spatial ability and numerical ability, which are required over a whole range of school subjects. Thus mathematical ability might be a particular hybrid drawn from a number of group factors but, if so, would be difficult to identify. Vernon (1950) reported research that showed 'common elements in attainments at different branches of mathematics, . . . a small mathematical factor, . . . a tendency for verbal ability to correlate negatively with mathematical . . . while spatial ability [correlated with] geometry only'. Wrigley (1963) was more positive in writing 'it appears that the evidence is overwhelmingly in favour of a group factor for mathematical ability, over and above "g", the general factor'. Nevertheless, 'g' was found to be very significant, thus 'high intelligence is the prime requisite for high mathematical ability'. Bell, Costello and Küchemann (1983) used Wrigley's work in claiming 'the relative independence of computational achievement is generally established', but also stated that 'not all factorial studies yield the same set of components [factors]'. The latter statement reflects the real problem, namely that factor-analytic studies have not led us to a full understanding of mathematical ability. There have been some benefits, however, for example the indication that verbal ability tends to correlate negatively with mathematical ability warns us of the danger of attempting to assess the potential of pupils on the basis of verbal tests alone—a practice still used in some schools.

Guilford (1959) used factor analysis to develop a unique approach to the study of human abilities. He proposed a cubical model of the intellect, with the three axes defining different kinds of classification (see Figure 7.1). One axis comprised the five major groups of abilities which he claimed were suggested by factor-analytic studies, and these were cognition, memory, convergent thinking, divergent thinking and evaluation. These Guilford termed 'operations'. A second axis was concerned with categories of content, namely figural, symbolic and semantic, where figural implied 'concrete material perceived through the senses', symbolic implied symbols of all kinds including letters and digits, and semantic implied verbal meanings or ideas. The third axis classified six kinds of products when a certain operation was applied to particular content, and these were units, classes, relations, systems, transformations

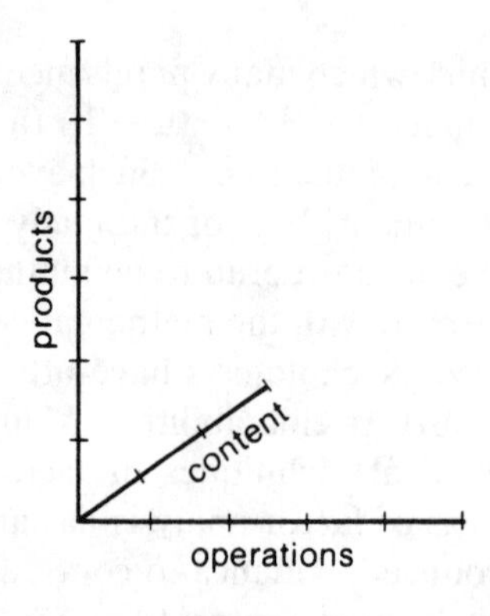

Figure 7.1

and implications, which he claimed factor analysis had suggested were 'the only fundamental kinds of products that we know'. The outcome was a 5 × 3 × 6 cubical arrangement, consisting of 90 smaller cubes or cells, each defining a particular ability in terms of a single operation on a single content type and a single product. Guilford was then able to demonstrate many of these particular abilities through sample test items. Examples of cognition/semantic/systems items in the mathematical domain can be provided, according to Guilford, by tests of 'necessary mathematical operations' such as this arithmetic reasoning item from CSMS (Brown, 1981a):

> A shop makes sandwiches. You can choose from 3 sorts of bread and 6 sorts of filling. How do you work out how many different sandwiches you could choose?
>
> 3×6 $6-3$ $6+3$ $3-6$
> $18\div3$ $6\div3$ 6×3 $3+3$

Examples of convergent and divergent items have been given in an earlier section of this chapter, though it is not easy to know which content and products these correspond to without careful reference to Guilford (1959). The divergent item concerned with making up equations from B − C = D and Z = A + D, also quoted earlier, is an example of a divergent/symbolic/implications item.

The implications of Guilford's model is that there are 90 distinct abilities. On this basis it is very difficult to talk of an entity such as mathematical ability. It is not impossible to think of this in terms of assembling a collection of appropriate cells and grouping these together as 'mathematical ability', though it is doubtful if this is a practical proposition. Guilford's subsequent discussion of mathematical ability reverts to a broader suggestion that it involves largely symbolic abilities, except that 'some aspects, such as geometry, have strong figural involvement', and semantic abilities are 'important in all courses where the learning of facts and ideas is essential'! Although Guilford's model of the 'three faces of intellect' is well known and is highly ingenious, it does not contribute much to the identification of the nature of mathematical ability.

More recent studies of mathematical ability based on factor analysis have been carried out by Furneaux and Rees. Reporting on two studies in a paper published in

1978 they claimed evidence to support strongly 'the view that there is a "mathematical ability" factor independent of "g"'. Earlier studies (Rees, 1974) had indicated that there was a core of mathematical test items which all groups of students found to be more difficult than other items. Factor analysis techniques subsequently led to the isolation of two relatively distinct types of mathematical ability (Rees, 1981). One type of ability, referred to as the 'g-factor', was found to be associated with more routine tasks and was dependent only on instrumental understanding (Skemp, 1976). The other type of ability was found to be related to making valid inferences, and was more dependent on relational understanding. Tasks which involved inference were more difficult for students than those requiring only the 'g-factor'. In considering implications for teaching, Rees suggested, perhaps controversially, that very able pupils should be positively encouraged to develop inferential powers whilst average and less able pupils should concentrate on intellectual development via more instrumental approaches, with the possibility that relational understanding might develop in some domains. It was not possible to be certain to what extent the inferential factor represents a specific mathematical ability.

The existence of different forms of high mathematical ability (Hadamard, 1945) together with the elusiveness of a single mathematical ability as revealed by factor analysis suggests that mathematical ability can take many forms, each form derived from a different mix of other abilities. These other abilities presumably include numerical ability, spatial ability, verbal and non-verbal reasoning, convergent and divergent thinking abilities, and so on. One particular ability which has attracted considerable research attention is spatial ability, which is now considered separately.

SPATIAL ABILITY

Learning mathematics involves the pupil with pictures, diagrams, graphs and visual presentations and representations of a wide variety of forms. One particular problem is the two-dimensional representation of three-dimensional objects. Thinking about three-dimensional objects is not particularly easy, unless the object itself is present. A well-known three-dimensional spatial problem involves a painted cube which is sliced across all three dimensions so as to produce smaller, congruent cubes. Two equally spaced slices across each dimension produces 27 smaller cubes. How many of these smaller cubes would have three painted faces, two painted faces, one painted face and no painted faces, respectively? What would be the numbers of cubes with 3, 2, 1, 0 painted faces if the original cube was dissected into 64 smaller cubes, 125 smaller cubes and so on? Would all pupils of high mathematical ability and sufficient maturity be able to visualize the effect of the dissecting? The evidence of differing abilities amongst great mathematicians would suggest not. Would some pupils who were not noticeably of high mathematical ability be able to solve the problem? One would suspect they would, particularly if spatial ability is a distinct ability and not simply one facet of mathematical ability. Spatial ability is certainly not linked only with mathematics, as examination questions in general studies papers indicate! The situation is even more complicated in that it might not be helpful to assume that spatial ability and ability to visualize are the same thing.

The anecdotal evidence of Hadamard exposed his own love of geometry and his ability to visualize, but it also cited Hermite's hatred of geometry. Krutetskii (1976), writing about school pupils who were mathematically very able, reported that some favoured spatial or geometrical thinking whilst others did not. Walkup (1965) provided further evidence of the capacity which certain people have to visualize in circumstances which do not at first sight appear to be conducive to visualization. Walkup hypothesized that some creative people have developed the ability to visualize in the area in which they are innovative. Thus Faraday could 'see' the electrical and magnetic lines of force, Kekulé could visualize the benzene ring as like a snake biting its tail, and Einstein believed that thought was a matter of dealing with mechanical images and was not concerned with words at all.

Smith (1964) carried out extensive studies into spatial ability which ultimately led him to conclude that spatial ability was a key component of mathematical ability. He also considered that the relationship between spatial ability and the cognitive trait which has become known as field-independence was a very strong one. The field-dependence/independence spectrum of cognitive style has been extensively researched, notably by Witkin *et al.* (1977). Such research has attempted to assess the extent to which the field which surrounds a situation influences perception. Examples of research studies include the documenting of reactions of people attempting to sit upright within a room which is tilted and also include documenting attempts to pick out a particular shape within a complex figure (see Figure 7.2). Towards one end of

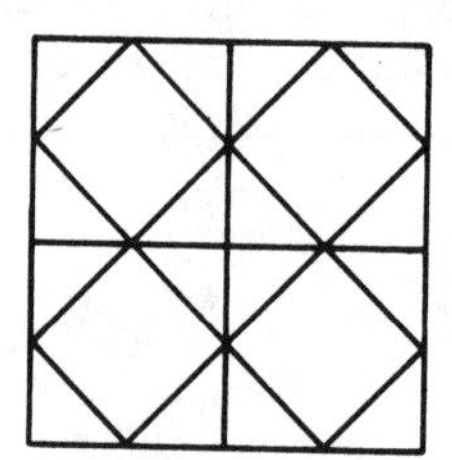

Figure 7.2

the spectrum those who are field-dependent have difficulty, for example in finding the shape within the complex figure or in finding true vertical in a tilted room. Those near the other end of the spectrum, the field-independent people, are able to ignore the confusion created by the surrounding field. Smith's perceived association between spatial ability and the field-dependence/independence issue is reflected in the inclusion of tests like the embedded figures tests in the catalogue of spatial ability tests (Eliot and Smith, 1983).

Bruner (1973), in stating 'I don't think we have begun to scratch the surface of training in visualization', was clearly raising the issue of the extent to which spatial ability may be enhanced through teaching. Mitchelmore (1980) suggested that measured differences in three-dimensional drawing ability between American, English and West Indian children were the results of differences in teaching approach: 'English teachers tend to have a more informal approach to geometry, to use more manipulative materials in teaching arithmetic at the elementary level and to use diagrams more freely at both secondary and tertiary levels.' The potential value of manipulative materials was also suggested by Bishop (1973) who found that children who have used such materials extensively tended to perform better on spatial ability tests than children from schools where such materials were hardly used. A recent issue concerning spatial ability relates to the widespread belief in mathematics-learning that drawing a diagram assists thinking. Indeed, one stage of the problem-solving process described by Polya included drawing a diagram. Perhaps it is the case that such a tactic only assists those whose ability includes geometric ability. Perhaps a diagram is a hindrance to those with limited spatial ability. A variety of issues concerned with spatial ability is discussed in Bishop (1980); this paper also includes a comprehensive review of the literature.

The possibility that there might be a spectrum rather than a dichotomy of abilities was raised in connection with field-dependence/independence. It is also possible, perhaps even likely, that all human abilities are similar. If there is a spectrum for every facet of ability and each one of us possesses a unique combination of levels of ability it should be no surprise that the study of mathematical ability, and of spatial ability, is so complex. There is certainly no evidence of one kind of mathematical ability which is the same for all mathematicians. Nor does research confirm that spatial ability is *the* most vital component of mathematical ability.

Additional insight into spatial ability has been obtained through medical research into the functions of different parts of the brain. In simple terms, the brain consists of a left hemisphere, a right hemisphere and a region linking the two hemispheres. A great deal is now known about the functions of various regions of the brain in terms of motor and intellectual skills (see, for example, Springer and Deutsch, 1981). Most of this knowledge is not relevant to spatial ability, but what is important is that the two hemispheres generally perform very distinct functions. The left hemisphere controls language and speech and excels in performing sequential tasks, logical reasoning and analysis. The right hemisphere processes stimuli as a whole structure (cf. Gestalt psychology) and processes images rather than words. A complex shape is seen as a whole by the right hemisphere whilst the left hemisphere analyses the parts separately. This brief outline is an over-simplification for, as might be expected, we are not all exactly the same. In broad general terms, however, it is true to say that the right hemisphere controls spatial ability. There has been the suggestion that school education tends to concentrate on developing those abilities controlled by the left hemisphere, whilst the right hemisphere is comparatively neglected. At the moment, we do not appear to know how to act on the knowledge we have about the link between abilities and the hemispheres of the brain in terms of providing appropriate learning experiences.

GENDER-RELATED DIFFERENCES

Evidence from around the world that there are sex-related differences in mathematical ability is not consistent. Differences in attainment in mathematics, as measured by public examination scores, are well documented for Britain and for many other countries. In Britain, little difference has been reported at the primary school level, though the Assessment of Performance Unit (APU) *Primary Survey* (1982a), in commenting on sex differences at age 11, recorded that what differences there were 'appear to foreshadow the main areas of mathematics where the differences are larger some five years later'. Leder (1985) also pointed to 'few consistent sex-related differences . . . at the primary school level'. In many countries, the post-primary pattern is the same; more boys than girls succeed in public examinations taken around the age of 16, many more boys than girls choose mathematics as one of their specialist subjects, and comparatively few females have, in the past, taken up employment directly related to mathematics or dependent on qualifications in mathematics. In the Soviet Union, on the other hand, it has been claimed that such marked differences are not evident. Krutetskii (1976) concluded from his studies of the mathematical ability of boys and girls that there was no clear evidence of any difference. Suydam and Weaver (1977) reported that, in the USA, sex differences do not appear to exist in the ability to solve arithmetic word problems.

Reasons for differences in attainment have been investigated from a variety of different standpoints including biological, psychological and sociological. The best documented conclusions have emerged in relation to the domain of societal attitudes and expectations. There is a strong indication that, in a variety of ways, girls have been consistently discriminated against in terms of mathematical education. This is clearly a serious issue, particularly since qualified mathematicians are permanently in short supply, and it was considered important enough to warrant a separate appendix in the Cockcroft Report (Shuard, 1982a). Sociological factors are not disputed, but the problem is whether there are other factors at work in addition to discrimination, and it is here that research evidence is particularly inconsistent.

The influences of society and from the environment which might affect the mathematical development of girls are varied. There have always been differences in the kinds of toys given to girls and boys and, indeed, in the kinds of games and activities encouraged. There have always been differences in parental expectations and desires which may even have led to differences in the pressures exerted within the home. The usefulness and value of mathematics have always generally been considered to be in other school subjects also regarded as boys' subjects and in careers which have been viewed by society as male occupations. Russell (1983), actually, has drawn attention to the fact that pressures might work equally unfairly against both sexes in that girls are not encouraged to opt for mathematical studies whereas boys are, even when their ability and interest in the subject have been, at best, barely adequate. Boys often opt for mathematics because it is expected of them and not because they enjoy the subject. Society always appears to have conveyed the message that mathematics is a male subject and that certain other subjects are female subjects. Peer group pressures therefore add to the difficulties faced by girls when choosing subjects in a mixed school. Russell also drew attention to the ways boys and girls regard themselves in relation to mathematical ability. Girls tend to underestimate their potential whereas boys tend to overestimate. Boys display confidence about their ability in mathematics which is sometimes not justifiable whilst girls, perhaps with better test results, display unjustifiable anxiety.

There might be other difficulties for girls created by the school environment. Most mathematics teachers are men. Many textbooks have, presumably unwittingly, insinuated a male image into mathematics. Some books have written men and boys into the text and exercises and have omitted women and girls almost completely. Many authors, perhaps because they themselves were male, have placed mathematics within contexts which were of much greater interest to males than to females. Internal school organization has not always allowed girls to combine the study of mathematics with their other choices, and girls in girls' schools have sometimes been guided away from mathematics because of timetabling difficulties. Teachers have been shown to interact in the classroom much more with boys than with girls, have paid more attention to boys, have given more positive encouragement to boys, have allowed boys to gain their attention simply because the boys clamoured more than the girls. Some research has suggested that girls achieve more in mathematics in a single-sex school than they do in a mixed school. Boys, on the other hand, tend to perform better in a mixed school than in a single-sex school! Such is the range and variety of influences which might result in differences in performance in mathematics that it is easy to appreciate the summary by Leder (1985): 'Sex differences possibly due to biological constraints are dwarfed by the far greater pressures imposed by social and cultural stereotypes about cognitive skills and occupations.' Nevertheless, the literature has also hinted at possible biological constraints and attention needs to be paid to these too.

In a discussion of differences in intelligence and special abilities between males and females Hutt (1972) clearly accepted that there were factors which originated from the biological and psychological domains. The main differences raised were as follows. First, scores obtained from applying tests designed to measure overall intelligence or ability consistently produced different distributions for males and females. The scores for males tended to spread more widely across the range whilst the scores for females were more clustered around the mean. The difference was not a large one, but there was a tendency for males to predominate in both extremes, the most able and the least able. Secondly, males were said to excel in spatial ability whilst females excelled in verbal ability. Field-dependence/independence tests had also produced evidence that males were more inclined to independence than were females. Thirdly, females were clearly superior in both manual dexterity and in rote learning ability whereas tests of divergent thinking tended to produce higher scores for males. In all cases the differences between the sexes were small and one must be careful in drawing conclusions. In such matters, differences between the extreme performances for either sex are always likely to be huge in comparison with any difference between the sexes. And none of the above results has been unanimously accepted as convincing in respect of innate differences as opposed to differences produced by differential experiences.

Benbow and Stanley (1980) carried out research involving intellectually gifted pupils. Their results, which were disputed by other researchers, suggested 'that sex differences in achievement in and attitude toward mathematics result from superior male mathematical ability, which may in turn be related to greater male ability in spatial tasks'. They did accept the clear and massive influence of socialization as a determinant of differences in performance in mathematics but did not accept that such social factors accounted for all of the differences. Of course, spatial ability is itself not necessarily vital to the whole range of mathematical activity. The debate about the relative contributions of environmental and social factors as opposed to innate characteristics seems likely to continue well into the future!

An interesting feature of the measured differences in performance is that it is only in certain mathematical topics that boys score very much higher than girls. The APU *Secondary Survey* (1982b) stated: 'In all three surveys, the mean scores of the boys have been higher than those of the girls in every sub-category, with only one exception (modern algebra).' The greatest differences were recorded in the topics of mensuration, rate and ratio, descriptive geometry and unit measures. Wood (1977) observed similar differences. Drawing from the examination scripts of boys and girls educated in the same schools he discovered that the superiority of boys was most marked in items of two types, one of which was concerned with ratio (scaling, moving between different orders of magnitude, moving between different units of measurement). Wood drew a parallel with the common core of difficulty discovered by Rees (1974) and claimed that the underlying difficulty was a 'comparison factor', a factor which was basically the scaling up and down which is so important in coming to an understanding of metric proportionality. It was no surprise that fractions were found to be more difficult for girls than for boys. This is a difficult topic area for both sexes but Wood asked whether this was a major source of the difference between the sexes? Perhaps a concentration of effort 'on fractions, proportion and, more generally, comparison factors' is what is required to compensate for the difference between girls and boys.

The other topic area observed by Wood as resulting in sex differences was geometry, also noted by the APU (1982b). The alleged difference in spatial ability has been mentioned earlier, and was also referred to by Wood in the statement: 'Girls' weakness at spatial visualization is, by now, well documented, and it is generally known that genetic causes are suspected', and in, 'Girls are notoriously poor at solid geometry problems.' Spatial ability, as we have seen, is not required in all of mathematics, only in certain aspects, and many famous mathematicians have not felt themselves to be at all capable in geometry. However, if we accept that spatial ability is comparatively weaker in girls than in boys, this is a major handicap to a study of some parts of mathematics. The greatest sex difference in mathematical performance found by Wood appeared in the results to a question which 'was an almost pure measure of ability to visualize in three dimensions'. Fennema and Tartre (1985), in reporting on a longitudinal study, confirmed that there was a difference between girls and boys in respect of spatial visualization skills, but it was small. They did agree that 'low spatial visualization skill may be more debilitating to girls' mathematical problem-solving than to boys''. It might also be debilitating in respect of attitude to mathematics. It should be pointed out, however, that evidence from other cultures does not always confirm that girls have weaknesses in spatial visualization.

Suggestions of comparatively poorer spatial ability in girls, and comparatively better verbal ability, have led to considerations of brain differences. The left hemisphere of the brain controls verbal abilities whilst the right hemisphere controls spatial abilities. Are the two hemispheres differently developed in girls and boys? At the present time research does not confirm that such a difference exists, and therefore cannot provide an alternative to the hypothesis that the differences in attainment are a product of environmental and social influences which perhaps determine the development of the various mental faculties within the brain. Other genetically based theories have also been proposed to explain the observable differences between girls and boys in terms of mathematical attainment. None is considered to provide convincing evidence that it is genetic factors which give rise to the differences.

PREFERENCES AND ATTITUDES

Hudson (1966) drew attention to the possibility that a liking for mathematics stemmed from preferred styles of study. Mathematics does not involve the learner in revealing emotions or opinions to others and hardly involves, of absolute necessity, any communication with others. The fact that mathematics provides 'a beautiful safe haven from the fears and anxieties of life' (Caldwell, 1972) is attractive to some. At the same time, private emotional reaction to the beauty or elegance of mathematical ideas and results is not ruled out. On the other hand the fact that 'maths is just a matter of facts being hammered into you . . . it's not a subject you can humanize' (Russell, 1983) is a deterrent to others. Russell also showed that pupils often perceive the mathematics classroom as being a place for competition, which is attractive to some and not to others. A competitive atmosphere can act as an incentive, particularly for successful pupils; for less successful pupils a negative attitude to mathematics is the outcome. Hardy (1940) attributed the development of his own deep interest in mathematics, at least in part, to the discovery that he was very successful in competition with other pupils. The security of the traditional method of teaching mathematics via exposition by the teacher followed by practice is attractive to some. The fact that responses or solutions are right or wrong, and discussion of the degree of rightness or wrongness is not appropriate, is liked by some pupils. Other pupils thrive on discussion, or wish to express their personality, and do not find that mathematics allows this. Preferences are part of the issue of individual differences and might exert a great impact in terms of why some pupils achieve more than others.

In terms of the totality of educational research, comparatively little work has been carried out in the domain of preferences and attitudes in learning mathematics. Earlier studies include those by Biggs (1962) and Husen (1967). The APU (1982a) found that the relationship between attitude and performance in mathematics at age 11 was 'surprisingly weak'. Boys demonstrated greater confidence in their own mathematical ability than did girls, and this was also reflected at age 15 (APU, 1982b). Mathematics was believed to be important by a majority of 15-year-old pupils and there was correlation between ratings of usefulness and interest.

Interesting results were obtained by Kempa and McGough (1977) in a study of attitudes to mathematics amongst sixth-form students. Differences in curricula did not appear to have resulted in differences in attitude. Curriculum innovators had often hoped that the modern syllabuses, introduced from the early 1960s onwards, would result in greater interest in mathematics. A curriculum is, however, more than just a syllabus. It is possible that teachers using newer syllabus material were not as confident as those using only the old familiar material. Some variables in educational research are very difficult to hold constant. Kempa and McGough also claimed that the research revealed that students' views about the difficulty of mathematics did not appear to have been a major determinant of whether students chose mathematics in the sixth form. Much more important were the perceived usefulness of mathematics and a liking for the subject. In terms of cognitive preferences, Kempa and McGough found that 'the preference for the symbolic mode of expressing mathematical information is seen to increase steadily with increasing mathematical bias of the students'. The three modes presented as alternatives were symbolic, graphical and verbal. Students with an arts bias tended to prefer the verbal communication mode.

Major studies of cognitive preference include those concerned with the field-dependence/independence issue, and these studies have already been referred to in connection with spatial ability. Bruner, Goodnow and Austin (1956) have suggested preferences in problem-solving methods between 'focusers' and 'scanners'. Thus, in looking for a relationship between items the 'focusers' would extract as much information as possible from the first item and then use the information as a basis for comparison and amendment in focusing on the other items in turn. 'Scanners', on the other hand, would select only one property and then scan all items, and would proceed by scanning all items for other properties. Under pressure of time, focusing was found to be the more effective method. Whilst such results might not be particularly important for mathematics-learning, they are indicative of the kind of differences which can occur in preferred ways of working.

It is important to realize that a decision to study mathematics does not imply a positive liking for the subject. Russell (1983) found many sixth-form boys studying mathematics who did not like the subject. They had opted for mathematics because they considered it a useful subject. Girls, on the whole, did not perceive mathematics to be all that useful, and there was evidence that it was largely those girls who really did enjoy mathematics who continued with the subject into the sixth-form. Mathematics was considered to be a high status subject, particularly by boys, but this does not of itself imply liking. The attitude of many girls to mathematics appeared to deteriorate steadily through the years of secondary schooling, alongside the growth of self-consciousness about errors and difficulties. There was some evidence from Russell's research that a good relationship with the teacher was more important for girls than it was for boys. A particular problem was that certain topics in mathematics were considered irrelevant by pupils. The teacher perhaps needs to make efforts to explain why certain topics are included, but there may be other topics which defy justification!

Mathematics teachers take for granted the acceptability of question and answer situations. There is never any deliberate intention to reveal the inadequacies of particular pupils to peers but that is precisely what can happen, and there is evidence that pupils can find the situation embarrassing and unacceptable. Holt (1969) has provided anecdotal evidence of the strategies pupils use to cope with the question and answer situation in class, and has also referred to cultures where such a situation—namely one in which individuals might be laid open to ridicule—would be completely unacceptable. Clearly, pupils who feel that they are being embarrassed in this way will develop a negative attitude to mathematics. They might become extremely anxious and hence, in Scheerer's sense of over-motivation (see Chapter 6), be unable to produce their best work. The whole subject of anxiety about mathematics has been comprehensively discussed by Buxton (1981).

SUGGESTIONS FOR FURTHER READING

Buxton, L. (1981) *Do You Panic About Maths?* London: Heinemann.

Fox, L. H., Brody, L. and Tobin, D. (1980) *Women and the Mathematical Mystique*. Baltimore: Johns Hopkins University Press.

Hadamard, J. (1945) *The Psychology of Invention in the Mathematical Field.* Princeton: Princeton University Press.

Hudson, L. (1966) *Contrary Imaginations*. Harmondsworth: Penguin Books.

Krutetskii, V. A. (1976) *The Psychology of Mathematical Abilities in Schoolchildren*. Chicago: University of Chicago Press.

Springer, S. P. and Deutsch, G. (1981) *Left Brain, Right Brain*. San Francisco: Freeman.

QUESTIONS FOR DISCUSSION

1 What should mathematics teachers be doing to provide opportunities for divergent thinking?
2 How should we provide for the very able pupil in mathematics lessons?
3 How should we ensure that neither girls nor boys are disadvantaged in mathematics lessons?
4 How should mathematics be taught so as to foster a positive attitude and prevent anxiety?

Chapter 8

Does Language Interfere with Mathematics-Learning?

ISSUES OF LANGUAGE

The fortnightly mental arithmetic test for a particular child regularly included a question of the type:

> What is the difference between 47 and 23?

This particular child thought the question rather odd, but nevertheless answered it, as follows:

> One of the numbers is bigger than the other.

When the test papers were returned the answer was marked wrong. Such was the fear which the teacher generated in the pupils that there was no question of going to ask the teacher why it was wrong. Better to try again next time and see what happened. Along came the next test, so our pupil tried an answer of this kind:

> One number contains a 4 and a 7 but the other number doesn't.

Naturally, this was also marked wrong. Next time, with growing desperation, our pupil tried again:

> One number is about twice the other.

The saga continued over many tests until our pupil asked a friend what the correct answer was, and why.

The story is a true one. It illustrates a relatively trivial but very common problem which pupils experience in learning any subject, including mathematics. Problems of language are obviously not unique to children of any particular country; the following example has been taken from the May 1964 *Arithmetic Teacher*:

> A . . . kindergarten teacher drew a triangle, a square and a rectangle on the blackboard and explained each to her pupils. One little girl went home, drew the symbols and told her parents: 'This is a triangle . . . this is a square and this is a crashed angle.'

Such stories are amusing to adults but they are often not the slightest bit funny for the pupil concerned. Obstacles can be placed in the path of children which have little to do with mathematical ideas but which are created because of problems of language.

There are many aspects of the issue of language and mathematics which might affect learning (see Austin and Howson, 1979). Anecdotes about children experiencing difficulties because they do not understand the words are not hard to find, but they do introduce the issue of mathematical vocabulary. Even if the vocabulary is appropriate there might be problems because children do not always interpret statements literally, but sometimes appear to change the meaning into what they think the teacher intended to say. The special symbols of mathematics, as an extension to the language of the mathematics classroom, cause additional problems. Reading mathematics is different from reading literature, or even from reading texts in other subjects. Learning mathematics in a second language can present difficulties. The place of talk in the classroom and the use of discussion between teacher and pupil, and between pupil and pupil, demands careful thought. The relationship between mathematics-learning and language development is clearly crucial. The extent to which the acquisition or formation of concepts in the mind of the learner depends on the use of appropriate language is an important issue.

Barnes has argued that teachers are often aware of problems of form but are less likely to show concern about problems of use. It is not the particular terms used which are critical, it is whether the underlying concepts and processes are being communicated, whether meaning is being conveyed. In short, 'There is a danger of reducing to lists of "difficult words" and to "readability measures" what is a more complex interplay . . . It is access to one another's meanings that matters in teaching' (Barnes, 1985). Skemp (1982) used the terms 'deep structures' and 'surface structures' to draw attention to the two levels, namely, the ideas of mathematics which we wish to communicate and the language and symbol systems which represent the ideas and which we use to transmit the meaning. It is important to be aware of potential problems of both form and use; we must know that problems might arise at the surface level of transmission and at the deeper level of meaning.

THE VOCABULARY OF MATHEMATICS

The boxed summary on the next page, taken from a mathematics textbook (with some alteration to the technical terms!), probably does not convey much meaning. To a child who did not grasp the technical terms during the two or three lessons in which

> *Summary:*
>
> A danding is a kambon if each trotick of the squidgment has only one ploud, for example the danding which dands each number onto its smallest lume tombage is a kambon.

they were introduced, or (in the case of two of the words) revised, the summary does not help. To a child who missed the lessons through illness the summary is meaningless. There is a very considerable vocabulary of mathematics in terms of number of words. Bell (1970) listed a basic vocabulary of some 365 words, in common use outside of mathematics as well as within, which even our slowest learners need to comprehend in dealing with the six elementary mathematical topics of quantity, measurement, time, money, position and natural number. Beyond that basic vocabulary, most children would be expected to learn the meaning of 100 or so new words in each year of mathematics lessons. This mathematical vocabulary ranges from simple words like 'find' and 'sort' to more specialized words like 'bilateral' and 'quadratic'.

Although there are serious problems in terms of the extent of the vocabulary of mathematics, there are additional problems when particular words carry a mathematical meaning which is different from the usual everyday meaning. One of the words in the summary example above is a replacement for the word 'relation'. A mathematical relation is a set of ordered pairs, but the ordinary use of the same word suggests a member of the extended family, and bears no resemblance to the mathematical idea. It is possible to use the ordinary meaning of 'relation' to introduce the mathematical idea but, given what we are aiming at, is that a help or a hindrance? Is it possible that, for some children, it is misleading to suggest that mathematical relations are a simple extension of the idea of family relations? As we have seen, the word 'difference' has a mathematical usage which is very specific and which needs to be made clear to pupils. There are, of course, many words which are wholly specific to mathematics and about which there should be no confusion with an everyday meaning, for example 'numerator', 'isosceles' and 'hypotenuse'. It is interesting that the word 'denominator' now occurs in certain everyday situations, perhaps indicating that one of the many changes which are constantly taking place in language usage is that technical terms are appropriated and redefined for ordinary, non-technical use. There must be many other words which were in use in everyday speech before mathematicians adopted them and assigned technical meanings; perhaps, for example, 'field', 'group' and 'root'. Examples of this two-way traffic must be numerous, but it is not the exact origins which are important, it is the learning problems which are created. Stories of children who think that 'volume' is merely a knob on the television or radio set (set!) or who think 'axes' are only for chopping with, or who believe that a 'revolution' is what happens when a government is violently overthrown are too frequent for us not to pay attention to the problem.

Some words are particularly difficult for children. It is very questionable whether more than just a small minority of our pupils ever distinguish the mathematical meaning of 'similar' from its everyday meaning. The particular problem here is that the two meanings are not far apart, the distinction is quite a subtle one. Perhaps the word 'relation' also comes into the same category as 'similar'. Another interesting example is 'segment', which sometimes appears to be used in everyday speech when the correct word is 'sector'. Some words are used with a different specialized meaning in other subjects, like 'chord'. A few words might be very difficult for children to accept, like 'vulgar' and 'improper', and yet others are odd, like 'odd' and 'real'. Clearly, it is very important, when a word like 'root' is introduced in a mathematics lesson, that the pupils are given the opportunity to come to terms with the particular mathematical meaning. Many words in a language have a number of meanings, and that might be a problem for younger children. Older children will have accepted this, but will still need time to get accustomed to any new usage. A salutary exercise for a teacher is to write down all the technical words used with a class throughout one year from the point of view of which category they fall into: unique to mathematics (and other studies when mathematics is being applied), not unique and subtly different in meaning, not unique and very different in meaning, odd or objectionable. Clearly, testing pupils, perhaps simply orally, will reveal the points of difficulty and will allow remediation.

Apart from technical terms which represent mathematical concepts, some of the more general instructions are not understood as well as might be imagined. Children have been known to think that 'evaluate' means 'change the value of', and even that 'tabulate' means 'take the tablets'. Frequently, the word 'simplify' does not really mean 'make simpler', it means 'carry out the mathematical process which has been demonstrated as being what is required'. An unusual problem of language is one that we create ourselves when we make a careful definition and then misuse the word. The most obvious example of this is when we describe axes on a Cartesian graph as 'horizontal' and 'vertical'. It cannot help the many children who are still struggling to understand what we mean by 'vertical' and 'horizontal' when we carelessly refer to one of the horizontal lines on the graph paper on their desk as 'vertical'. The word 'histogram' has a very precise meaning in mathematics which distinguishes it from a 'bar chart'. Unfortunately, there are situations in mathematics in which the boundary between the two ideas appears to be hazy and ill-defined (see Chapter 3) and this has led to misuse of the term 'histogram'. Another word which can create problems is 'chord'. There was a time when the word 'secant' was used to describe the line which was obtained when a chord was produced in both directions, but now we often attempt to use the word 'chord' for this extended line as well. We therefore have no idea what meaning is being registered in the mind of the child when we use the word 'chord', and this can be critical when it comes to introducing differentiation (Orton, 1983b). It is interesting that even the word 'produced' can cause problems, and many pupils cannot understand why we do not use 'extended'.

It might be that problems of vocabulary are considered to be fairly superficial within the whole issue of language and mathematics-learning, but it is nevertheless critical that such problems are not ignored in the hope that they will go away. In order to facilitate the learning of mathematical ideas it is important that children are given help with the language which they are going to be expected to use in discussing and generally processing those ideas. It is clearly necessary when a new idea and its surface representation are introduced that any words and symbols are written down as well as spoken, and that they are discussed. There are also a number of specific vocabulary activities which are well known to children through work in other subjects and through puzzle books. A simple word search based on the mathematical topic just completed can help with consolidating the words and their spellings. An elementary crossword can be used to attach the word to its meaning or to the way it might occur in mathematical art. Unscrambling words (e.g. UARQES), matching a list of words with a list of meanings or a set of pictures, and a mixture of unscrambling and matching, all have some value. Selecting the best description for a word from a number of alternatives is another possibility. Using lesson time for such activities might be considered a nuisance until one realizes that coping with the vocabulary is an important part of learning the ideas.

READING MATHEMATICS

Any mathematical textual material which is to be read by pupils must be 'readable'. It is not easy to define readability in completely explicit terms, but there is no doubt about what is meant; we mean that pupils should be able to learn what we intend them to learn without the language itself getting in the way. Thus, the lengths of words, the lengths of sentences, the particular words used and whether they form part of the vocabulary of the pupils may all be considered important. The readability of text has become of major concern in education in recent years, and a variety of different techniques has been proposed to enable teachers to carry out checks as to whether particular text is appropriate. The difficulty of defining readability seems to have been reflected in these proposals, however, because most of the techniques only incorporate a selection of the possible facets of readability. Thus the Dale–Chall formula is based only on percentage of words not included in a set list of common words and the average number of words in a sentence. The FOG formula is based only on the average number of words in a sentence and the percentage of words with three or more syllables. The Flesch formula is based on the average number of syllables per 100 words and the average number of words per sentence. The Fry procedure is based on the number of syllables and the number of sentences in a 100-word passage. And the 'cloze' procedure is based on the ability of the reader to fill in missing words in text. Such methods were clearly not developed with mathematical text in mind, nor are they necessarily applicable outside a particular region of the world. Thus, for example, a formula devised for use in the USA would not necessarily be as applicable in Britain. The various readability formulas have been reviewed in more detail in Shuard (1982b) and in Shuard and Rothery (1984).

The readability formulas outlined above, and others not mentioned, are generally not applicable to mathematics because mathematical text is peculiar in comparison with text in other subject areas. The text does not necessarily flow left to right, line after line. It is sometimes necessary to move in unusual directions and even to move about the page in order to refer to tables, graphs or diagrams. The text is also likely to contain certain non-alphabetical symbols, which may or may not be numbers. This complexity of mathematical texts led Kane, Byrne and Hater (1974) to devise a readability formula for use specifically with mathematical text. Unfortunately, not only is the formula extremely complex to apply, it might not give the same results outside the USA. In general, apart from drawing the attention of mathematics teachers to the important issue of readability, specific formulas have not proved to be of much value. It is essential that teachers of mathematics should assess, critically, the appropriateness of text for their pupils, but the general recommendation at the moment is that this should be achieved by means of 'informed judgements'. Clearly, it is important to have the information on which to base such a judgement.

It is possible to read a story or novel in English in a fairly superficial way, and yet still derive meaning, message and moral. It is even possible to use rapid reading techniques, perhaps skipping sentences or descriptive paragraphs which are clearly not crucial. Non-fiction cannot generally be read in a superficial way without losing detail that might be essential, and mathematical text comes into this category. Mathematical text generally cannot be read quickly, for every word might be crucial and every symbol essential in the extraction of meaning. In order to ensure that attention is focused on all parts of the text, interaction is important. The issue of interaction with text arose in Chapter 4 in a consideration of programmed learning. Such stimulus-response-based interactive texts are not currently popular for reasons which have nothing to do with the acknowledged importance of interaction. Many children learn mathematics today through the use of workcards which clearly necessitate interaction and which contain some explanatory text. The entire workcard scheme can be considered as a large-scale interactive text, similar in many ways to a programmed learning scheme. However the material is presented, though, whether textbook or workcard, some parts will demand interaction and other parts will not.

Shuard and Rothery (1984) described the main components of mathematical text as teaching, exercises, revision and testing. The last two, revision and testing, could be considered to be repetitions of teaching and exercises but presented in a different, perhaps abbreviated, format, and included for a different purpose. For many pupils, and in the interests of readability, it is necessary that the teaching sections should be quite short, or that exercises and teaching are interwoven in order to achieve interaction. However, having achieved interaction within the text, there is no guarantee that pupils will interact in the way intended. How do the pupils know that their responses are correct? How are they to be informed whether they reacted correctly or not? How do we ensure that pupils do not skip sections and resume reading at the next point where answers are provided? Many of the problems associated with interactive text are the same as some which were discussed in the programmed learning section in Chapter 4. The boxed extract from the Manchester Mathematics Group (1970) illustrates the difficulties indicated above. It is not easy to provide an interactive, readable text.

Other features of mathematical text not always found in textual material are graphs, tables and diagrams. These offer the advantage that they break up the text. However, it is essential that they are situated appropriately in relation to references to them within the text. It is also essential that pupils are compelled to interact with them and are assisted in extracting information from them. Worked examples are commonly found within mathematical text, and these can form an important reference for pupils when working alone. Clearly, such worked examples are of no value if they are ignored, both by pupils and teacher. Interaction may also be achieved through practical activities directed from within the text, and then it is helpful in stressing the importance of the practical work if the results of the activity are referred to in the subsequent development of ideas in the text. Before interaction of any kind can be achieved, however, the pupils must find the text sufficiently attractive. Variety in

Powers of 2

In the following sequence, consecutive numbers always bear the same relationship to each other. What numbers will fill the gaps:

2, 4, _, 16, 32, __, ___, 256, ____, 1024, _____, . . .

What would the 12th term of the sequence be? What is the connection between each term of the sequence and the next one?

Now write down the first 10 terms of this sequence (on a single line). Underneath each term write the number showing the place that term occupies in the sequence, e.g. 16 is the fourth term, so write 4 under 16. If you have a coloured pen or pencil, write the second row of numbers in a different colour. The second row of numbers also forms a sequence. What is the relation between successive terms in this sequence?

To avoid repetition, we shall call the two sets of elements of the two sequences B and I. Thus

B = {2, 4, 8 . . . 1024}

I = {1, 2, 3 . . . 10}

The table formed from the two sequences shows how the elements of B and I are related.

colour, in type style, in spacing and general layout all have a part to play. Pictures are valuable, in commanding attention, though it must be remembered that many adolescents are easily offended if they feel the attempts to brighten up the material are beneath them.

The correct and most appropriate forms and level of interaction are important in ensuring readability but so too is whether the meaning flows for the pupil. Shuard and Rothery (1984) have drawn attention to this issue and to the three types of meaning unit included within mathematical text. First, there are explicit statements, for example, 'In the following sequence, consecutive numbers always bear the same relationship to each other.' Secondly, there are statements or questions which demand activity from the pupil. This interaction ultimately provides information which is intended to be explicit, as the first kind of statement, for example, 'What numbers will fill the gaps:

2, 4, –, 16, 32, —, —, 256, ——, 1024, ——, . . .?'

Thirdly, there are gaps, meanings which are essential but which are not provided by either the first or second type of statement. It is necessary for the pupil to fill such gaps, either by inference from the text or by bringing knowledge from completely outside the text. An example of this from *Peak Mathematics 5* (1982) is given in the box.

Clearly, in this example, there is more to deciding on this 'same denominator' than writing down any number, and here the pupils either must infer that the required denominator is the product of the two given denominators or must bring knowledge about lowest common denominators or equivalent fractions from outside. Even then there is more for the pupils to fill in, for no mention has been made of numerators. In

To add fractions,
the denominators
must be the same.

$\frac{1}{3}+\frac{1}{2}$
$=\frac{2}{6}+\frac{3}{6}$
$=\frac{5}{6}$

Add these fractions.

1. $\frac{1}{4}+\frac{1}{3}$, etc.

many cases, the third type of meaning unit has been provided more explicitly on an earlier occasion, but not always.

Shuard and Rothery (1984) suggested a diagrammatic analysis of flow-of-meaning, using three different symbols for the three kinds of unit. The value of such an analysis to the teacher was summed up in:

> Pupils are not likely to notice many of the subtle points which the flow-of-meaning diagram brings out, but they may experience a feeling of general inability to follow the argument of the passage . . . Thus, a flow-of-meaning analysis may be useful in trying to understand the difficulties which pupils find in a particular passage, or in preparing a discussion lesson as an introduction to the work.

Unfortunately, the preparation of a flow-of-meaning diagram for even a short section of text may take considerable time.

In order to make an informed judgement about the suitability of text for particular pupils it therefore appears that there are many facets of the text to take into account. At the most superficial level there is the general attractiveness and appeal of the text. Within this more superficial level one might also consider the general style and layout in terms of variation of type and spacing, and the use of methods of highlighting. Appropriate summaries can also be helpful in fostering retention. Then there is the relationship between text to be read and sections to provide active involvement. A more detailed scrutiny then needs to take in vocabulary, length of words, and length and structure of sentences. Particular features of mathematical text like tables, graphs, diagrams and special symbols need to be inspected. The flow-of-meaning requires detailed analysis if it is to provide useful information. Finally, the exercises themselves need to be considered, for they too must flow, from the relatively elementary to the more difficult; they must be correctly sequenced and appropriately paced.

MATHEMATICAL SYMBOLS

In one sense, there is no difference between symbols which are specific to mathematics and other symbolic ways of representing ideas. The symbol '4' and the word 'four' convey the same message, only the form is different. In reading mathematical text it is necessary to be able to read those special mathematical symbols which are incorporated within the text for, just as a word represents an idea, so also 'without an idea attached, a symbol is empty, meaningless' (Skemp, 1971). In another sense, however, the special symbols of mathematics are different. To the experienced student of mathematics, the expression $(3 + 2) \times 4$ conveys a clear message but at the same time it is very difficult to dictate the expression to someone else. Some arrangements of symbols are not left to right, for example $\frac{3}{4}$. In fact $\frac{3}{4}$ is a special symbolic arrangement representing $3 \times \frac{1}{4}$ and is comparable in this respect to $3(x + 1)$. Sometimes it is necessary to read a collection of mathematical symbols before a clear message can be obtained, for example $\int_1^2 x^2\, dx$. Sometimes different structures of symbols are intended to convey the same meaning, for example $3 \div 4$ and $4\overline{)3}$, and at other times the same structure implies different meanings, for example 34 and $3x$.

Many problems occur for children in coming to terms with mathematical symbols and their corresponding words. In $5 + 3 = 8$, for example, it is common for the '+' to be read as 'add'. Yet the instruction 'add 5 and 3' would not be written as '+5 and 3'. The symbolic representations '+' and 'add' are not exactly interchangeable. In the same example '=' might be read as 'makes' at first, but then eventually as 'equals'. Yet 'makes' would not be considered an acceptable replacement for '=' in

$$\frac{d^2y}{dx^2} + 3\frac{dy}{dx} = 7.$$

The same symbol '=' might be read as 'leaves' in $5 - 3 = 2$. The instruction 'take 3 from 5', and even the more advanced equivalent 'subtract 3 from 5', requires reversal of order in conversion to $5 - 3$. The symbol '×' in 3×5 is read as 'times' by many pupils throughout their entire mathematical education and implies 3 lots of 5, but the secondary school teacher wants 'multiplied by' which conveys the idea of 5 lots of 3. The symbol '÷' is read by many pupils as 'shared by', despite the fact that $6 \div 3$ can represent the sharing of 6 items between 3 receivers and can also represent the number of groups of 3 items which can be obtained from 6. Division also presents another problem in that $6 \div 3$, $\frac{6}{3}$, 6/3, $3\overline{)6}$ and $3\underline{|6}$ are all found in mathematics texts as different ways of setting up the same process. These five alternatives, it will be noticed, contain the two numbers, 6 and 3, in two different orders. The order does not matter when recording addition and multiplication because of commutativity, but it does matter in subtraction and division. We should not be surprised that $8\overline{)4}$ frequently leads to the response, '2'!

Discrimination between very similar symbols is vital in mathematics. The early symbols '+' and '×' are very similar, as are '−' and '÷'. In the secondary school the very different meanings of $2x$ and x^2 cause many problems. In discussing the symbols of mathematics Skemp (1982) has used the idea of surface structures and deep structures. The surface structures are the form of the symbols and these are intended to convey meanings, which are the deep structures. The symbols 23, $2\frac{1}{2}$, and $2a$ all have the same surface structure, but their deep structures are very different. Some pupils eventually have to be able to discriminate between a wide range of symbols with similar surface structures, for example $5C_2$, ${}^5C_2$, $5C^2$, ΣC^2 and ΣC_2. Most individual mathematical symbols have separate meanings, but a few do not. The fact that δx is a single unit and not a product like $3x$ causes problems. In fact, in terms of elementary calculus, the situation is much worse (Orton, 1983b), for $\frac{\delta y}{\delta x}$ is a quotient, δy divided by δx, but $\frac{dy}{dx}$ is not a quotient, and needs to be regarded as a single entity. Once we have convinced students of this we can then feel free to start using $\frac{dy}{dx}$ as if it was a quotient! A discussion of the symbols of mathematics is included in Shuard and Rothery (1984).

From the point of view of understanding symbols in mathematics, Skemp (1982) made a number of suggestions. The critical problem was taken to be one of lack of understanding of the deep structures, thus the suggestions have similarities with those intended to encourage concept formation. First, the symbolism should only be introduced as the final stage of a learning sequence which is developed from physical or concrete embodiments of the concepts. Place-value provides a good example of this, and many teachers have always provided a variety of forms of equipment from straws and bundles of straws to the more sophisticated Dienes Multi-base Arithmetic Blocks (Dienes, 1960). Secondly, the mathematical ideas should be sequenced and presented so that assimilation to existing conceptual knowledge is eased and should not be presented as a discrete unit of mathematics which bears no relationship to work which has gone before. Thirdly, spoken language needs to be used for much longer, and pressure to convert to abbreviated symbolism should be resisted. Finally, transitional notations should be used to form a bridge to the condensed symbolism, for example children might be happy using 'area = length × breadth' when they are not ready to accept $A = lb$. The introduction of mathematical symbols too soon, without an adequate understanding of the deep structures, is a major cause of alienation: 'Algebra is . . . a source of considerable confusion and negative attitudes among pupils' (Cockcroft, 1982)

COMMUNICATING MEANING

Conveying meaning to pupils is the objective of teaching. This will not necessarily be achieved even when vocabulary is appropriate, symbols are understood and text is readable, for a whole variety of reasons including the very important one that pupils will sometimes place their own interpretation on what we say. When we ask our own children at home to put the knives and forks on the table we do not expect to find that every single knife and fork in the house has been set out. We expect our children to interpret our vague message for themselves and only set out as many knives and forks as will be needed. We expect the children to know what we mean. The communication of meaning frequently involves interpretation on the part of the receiver, and this should warn us that messages could frequently be given incorrect interpretations. In the school situation children do not always interpret our words in the way we intended. Donaldson (1978), who used the 'knives and forks' illustration, suggested:

> When a child interprets what we say to him his interpretation is influenced by at least three things . . . his knowledge of the language, his assessment of what we intend (as indicated by our non-linguistic behaviour), and the manner in which he would represent the physical situation to himself if we were not there at all.

The nature of the responses of young children to questions put by teachers persuaded Piaget that these responses were a consequence of their stage of development and of the egocentric nature of their thinking (see Chapter 5). Donaldson and others have conducted experiments which, they have claimed, suggest that the difficulty is, in part at least, a problem of the interpretation of language. In one particular experiment four toy garages and a number of toy cars were used. A toy panda then made judgements on the truth or falsity of certain statements, and the children had to inform the panda whether the judgement was correct or not. The statements were: (1) all the cars are in the garages, (2) all the garages have cars in them. The numbers of cars present was: (a) 3 in the first experiment, (b) 5 in the second experiment. The expected responses were therefore: (a1) true, (a2) false, (b1) false, and (b2) true, but these expected answers were not provided by all the children. The conclusion drawn was that, irrespective of the actual words used, the children were concentrating on whether all the garages were full. There was also the suggestion from some responses that children were interpreting 'all the cars' to mean 'all the cars which ought to be there', in just the same way that they have to interpret our instructions about setting the table.

Figure 8.1

One of Piaget's conservation tasks involved two sticks arranged first in exact alignment (Figure 8.1 (a)) and then secondly with the alignment destroyed but the sticks still parallel (Figure 8.1 (b)). The fact that many younger children, answering the question, 'Are the sticks the same length?', said 'Yes' for (a) and 'No' for (b) was taken by Piaget to imply that such children did not accept conservation. If, however, these children were interpreting the situation as one in which the experimenter very deliberately draws attention to a change, and if they did not pay much attention to the exact words and their meaning, it is reasonable to assume something other than lack of appreciation of conservation. Repetition, by Rose and Blank, of the experiment without the first stage of exact alignment (reported by Donaldson, 1978) produced fewer errors. In the original experiment the children were not, perhaps, paying attention to the language so much as to other cues provided. It is important to note that alternative experiments have not produced correct answers from all children; they have produced a different balance of responses which has been taken to suggest that children have not interpreted language in the way that was expected.

Such results have drawn attention to the whole relationship between language and learning, whatever the school subject, for 'behind words there is the independent grammar of thoughts' (Vygotsky, 1962). The fact that children can interpret what we say in a way that is different from what we expected is but one part of the relationship between language and learning. The Cockcroft Report (1982) included the recommendation that mathematics teaching should include opportunities for discussion between pupil and teacher and between pupil and pupil. Bruner (1966) declared that language was 'not only the medium of exchange but the instrument the learner can use in bringing order into the environment'. Language plays a vital role in learning in that 'it makes knowledge and thought processes readily available to introspection and revision' (Barnes, 1976). Thus egocentric speech, talking for the benefit of oneself, is important for young children for 'it serves mental orientation, conscious understanding; it helps in overcoming difficulties; it is . . . intimately and usefully connected with . . . thinking' (Vygotsky, 1962).

The relationship between language and thought has been the subject of debate by psychologists over many years. To Piaget, language was important but it did not play a central role in the growth of thinking. Language 'helps the child to organize, . . . experience and carry . . . thoughts with precision . . . [but] this can only be brought about by dialogue and discussion alongside action' (Lovell, 1971b). To Vygotsky, language played a far greater role in the growth of thinking, for egocentric speech 'soon becomes an instrument of thought in the proper sense—in seeking and planning the solution of a problem'. However, although there might be considerable difference in emphasis in these views, the relationship between language and learning cannot be ignored in mathematics-learning. One outcome of the language issue has been emphasis on the importance of allowing children to talk about their mathematics.

Clearly, allowing children to talk offers advantages to the teacher in that access is obtained to the thinking of the pupils. Traditionally, this access has been obtained through question and answer but, although this is valuable, there is considerable doubt as to whether it allows sufficient involvement for pupils. In any case, evidence provided by Holt (1969) has suggested that teachers do not always obtain worthwhile feedback from such so-called classroom discussion situations, because children possess many strategies to deceive. Real contact with each individual child is only occasional in the traditional question and answer routine, and assumptions are often made by the teacher on the basis of non-linguistic cues. Flanders (1970) suggested that a 'rule of two-thirds' is in operation in most classroom situations, in that, 'In the average classroom someone is talking for two-thirds of the time; two-thirds of the talk is teacher-talk, and two-thirds of the teacher-talk is direct influence.' Direct influence was defined as including expounding, giving instructions and exercising authority. Very little of most lessons consists of genuine discussion, and even teachers who have believed that they were conducting a discussion lesson have been very surprised and upset on hearing the transcript. If a teacher is to obtain the kind of access to a pupil's thinking which is desirable, a one-to-one situation is required for a much longer time than in the normal question and answer situation. Extended one-to-one interview situations are the norm in many research studies (see Chapter 2), but most teachers do not manage to find much time amongst all their other duties for this kind of contact with pupils. Discussion between pupil and pupil in a small group situation is much easier to achieve, but might require follow-up discussion with the teacher.

Other criticisms of teacher-led discussions have been provided by Sutton (1981), reporting on the work of Adelmann and Elliott. First, such discussion often allows only the teacher to ask the questions, so the questions are not necessarily those which the pupils would ask and for which the pupils feel they need an answer. The teacher's questions are almost inevitably carefully sequenced to lead towards a defined objective. Secondly, even when pupils are allowed to ask questions of the teacher, the response given effectively kills off the likelihood of contributions from other pupils at that point, and so does not necessarily solve the pupils' difficulties. Those points which are raised by pupils are reformulated in the words of the teacher, anyway. Thirdly, it is inevitable that the teacher will take up some contributions from pupils and will reject and subsequently ignore others, which might do nothing to help pupils whose contributions are not discussed. Finally, it is also inevitable that some contributions from pupils will be received in a complimentary manner by the teacher, clearly indicating that these were the correct answers, the ones the teacher wanted all along, and preventing other pupils from subsequently suggesting alternatives which might have highlighted their particular difficulties.

The real objective of discussion is to foster learning. The growth of relational understanding requires constant appraisal and development of existing knowledge structures in the light of new knowledge. Concepts, as we have seen, are not formed and learned only to remain permanently fixed, they continue to change and develop as new contexts emerge and are studied. Talk allows appraisal and development of ideas to occur. Most of us are unwilling to concede that we talk to ourselves. Young children are not at all inhibited in this way, and engage constantly in egocentric speech. Older children, like adults, are not willing to be found talking aloud unless to others. Piaget believed that egocentric speech was a feature of a particular stage of growth and that it eventually disappeared, but Vygotsky believed that egocentric speech turned into inner speech, and in that form remained a feature of how ideas were manipulated. Inner speech, however, is not always sufficient, even for adults, and conversation often includes considerable periods of time when another adult is being used virtually as a sounding-board. Barnes (1985) stated:

> Verbalization is important because ideally it makes thought-processes open to conscious inspection and modification. It seems likely that verbalizing . . . aids the retrieval of schemes, their manipulation and combination, and the evaluation of their appropriateness.

Until recent times the majority of mathematics teaching has been convergent in nature, in that the objective has been to steer or pilot pupils towards learning a defined piece of knowledge. Only recently have more 'open-ended' situations been encouraged, with the possibility of divergence. Although discussion between pupils need not be thought of as necessarily leading to divergence, there is no doubt that discussion allows exploration of many ideas and allows informal hypotheses to be formed and tested. In terms of solving problems, a number of minds attacking the same situation in, inevitably, different ways ought to offer some advantages. Gagné and Smith (1962) produced some evidence that pupils who were encouraged to talk about what they were doing as they attempted to solve a mathematical problem were more successful than those for whom talk played little part. Wall (1965) commented:

> groups are more productive of hypotheses and therefore are likely to be more productive of solutions than single persons, though in fact they take more time. The solutions reached tend to have a higher quality . . . There also tends to be a higher level of criticism of the hypothesis and of the solutions in a group.

Teachers might be concerned with the suggestion that group methods take more time. In the long run, however, if ideas are grasped more thoroughly as a result of having been achieved with the help of discussion, time will be saved. So much of mathematics teaching seems to consist of teaching the same ideas again because pupils have not retained them. Clearly, for discussion to offer advantages to all concerned, every pupil must be actively involved in the sense of attending to the discussion in its entirety. Potential advantages of group discussion therefore might not materialize if the group is too large for all members to remain involved. Some pupils will 'sleep' if given the opportunity. Others will not be confident enough to make much contribution, particularly if there are dominant pupils in the group. However, all the evidence we have about using discussion between pupils to facilitate learning suggests that discussion is an important vehicle for sorting out ideas. Much more work still remains to be done to explore 'styles of mathematics teaching which will enable pupils to develop their mathematical understanding and thinking through varied use of language' (Torbe and Shuard, 1982). The current interest in classroom talk is illustrated by the recent publication of books such as Pimm (1987) and Brissenden (1988).

LANGUAGE, CULTURE AND MATHEMATICS

Communicating mathematical ideas so that the message is adequately understood is difficult enough when teacher and learner have a common first language, but the problem is more acute when their preferred languages differ. Many pupils, in most countries of the world, are expected to learn mathematics through the medium of a spoken and written language which is not the one used in the home. Mathematics is a very important subject in the primary curriculum, so the teaching of mathematics might have to commence in one language, only to change to another later. Whatever language is used for teaching purposes one would expect that pupils would have some knowledge of that language, but it might be a very restricted knowledge. Language is important not only for communicating but also because it facilitates thinking. The language used for thinking is most likely to be the first language, thus mathematics communicated in one language might need to be translated into another to allow thinking, and then would need to be translated back in order to converse with the teacher. Errors and misunderstandings might arise at any stage of this two-way inner translation process. In fact, Berry (1985) has contrasted the progress in mathematics of a group of university mathematics students in Botswana and a similar group of Chinese university students in Canada. The former group claimed they had to do all their thinking in English, because their own language did not facilitate mathematical proofs, and they did not find this easy. The Chinese students, on the other hand, claimed that they carried out their proofs in Chinese and then translated back to English, and that they were able to do this quite successfully. Berry certainly concludes that the more severe problems would be likely to lie with students trying to learn mathematics through the medium of an unfamiliar language which is very different from their own.

There are many problems created by the interaction between language and mathematical education and these were the subject of a major international (UNESCO) conference in 1974. Morris (1974) drew attention to the fact that those children who receive their mathematical education in the same language consistently throughout their school lives are fortunate; for a variety of reasons many do not. The variety of local languages in some countries necessitates that a national decision should be made about the teaching language. Ethiopia, for example, has 70 different languages and many more dialects, and Tanzania has some 120 languages. A common experience for many children is that they use a local language at home, learn at primary school in a regional or national language, and finally follow more advanced studies in one of the international languages. Some countries have taken the decision to adopt a bilingual approach. There is a variety of reasons why it is not appropriate to attempt to educate children throughout their school life in their local language, and there can also be problems with regional and national languages.

The first problem which often needs attention is lack of vocabulary and symbolism. Despite a reputation for mental arithmetic, the Yoruba (of Nigeria) had no symbols for the numerals nor for elementary mathematical operations. Despite the existence of symmetrical and octagonal constructions there were (in Ethiopia) no Amharic words for 'symmetry' and 'octagon'. There was some lack of precise equivalents for words denoting mathematical operations in Sinhalese. There was no word in Norwegian for 'power', so the same word as for 'force' was used, which is a horrific thought for those currently researching into students' understanding of mechanics. Major programmes of language enrichment have been undertaken around the world—in, for example, Malaysia, Indonesia, Tanzania and many other Asian and African countries. Often, words adopted have been taken from another language, but differences in association of word-form with pronunciation have led to different spellings, thus 'cube' in Malay has become 'kiub'. Sesotho, the language of Lesotho, is also the language for many people in South Africa, and because of this different words for the same concepts have been adopted in the two countries amongst people using the same language. Problems of lack of vocabulary can, of course, eventually be solved, but a more difficult problem arises when ideas do not exist.

The Yoruba traditionally compared weights by lifting by hand, and the ideas of weight and measures of weight were absent, whilst in Amharic ideas of, for example, negative number and square roots were new. The idea of a day in the Yoruba culture is that it lasts from dawn to dusk, not for 24 hours, and time is also measured from sunrise to sunset by those who speak Amharic. The ideas of zero and the empty set are very difficult to explain in Igbo (a Nigerian language) because of problems of the language representing slightly different ideas. Subsets are difficult to explain in Sinhalese and some common kinds of mathematical questions are rendered ridiculous, for example the translation of 'Are roses flowers?' is 'Are rose flowers flowers?'. The Yoruba idea of direction is imprecise, being based on directions of sunrise and sunset. Inclusive calculations in Yoruba and in other cultures lead to the translation of 'the day before yesterday' as 'three days back'. There are cultures, like the Oksapmin of New Guinea, where there is no concept of number base and body parts are used for counting. Even where the idea of base is part of the number system the base might not be ten (see Saxe and Posner, 1983), and thus translation of mathematics into a second language might create problems because the mathematical constructions are different. The Yoruba are said to have an unusually complex system involving base 20, but other bases exist, for example base four (the Huku of Uganda) and base fifteen (the Huli of New Guinea). The Dioulas (Ivory Coast) are able to identify commutativity in addition but not in multiplication because of asymmetry of linguistic construction, which hinders any concept that the multiplier and multiplicand may be exchanged. Morris (1974) has detailed many more examples of such difficulties across many cultures.

The problem may be even deeper than one of vocabulary and mathematical ideas, for there is also the suggestion that there is a problem caused by the 'distance' between the mother tongue and the language of instruction, which is also the language which has dictated the design of the curriculum. There is now the strong belief that the so-called Indo-European languages are 'close', but as a group they are far removed from, for example, the languages of many African countries. Berry (1985) has summarized these problems as:

> In general it is likely to be easier for a student to function effectively in a second language which is semantically and culturally close to his mother tongue than in one which is remote . . . [for] . . . the structure of a person's language has a determining influence on that person's cognitive processes . . . such as classification and recognition of equivalences—processes which are central to the understanding of mathematical concepts.

Clearly, if the problem is only that the language of instruction is not the learner's mother tongue, then it is necessary to provide remedial help of a linguistic nature. If, however, the problem is one of 'distance', and this problem can arise among unilinguals being taught in their own language, the appropriate remedial strategies are more likely to involve the mathematics rather than the language of instruction. There would seem to be a great need to develop mathematics curricula which enable and encourage students to think in their mother tongue.

Berry (1985), in recounting the difficulties faced by two children, Mothibi and Lefa, in their school situations in Botswana, suggested that school is a threatening place because 'mysterious tasks are assigned for no apparently useful reason'. The result is that disappointing progress is made in mathematics, and what progress is made is largely based on rote learning. Gay and Cole (1967) concluded from their research with the Kpelle (Nigeria) that there were no inherent difficulties about learning mathematics, it was simply that the content imposed by the curriculum did not make any sense within the Kpelle culture. All over the developing world one hears of disappointing mathematics results and of great concern about the very small numbers of pupils who show the expected level of mathematical competence, despite curricula which, though derived from Western curricula, have been meticulously translated so as to reflect the world of the children who will use the learning materials. Could there be more than a problem of language here? Anthropological and linguistic studies appear to indicate that language and culture are inseparable, so that no amount of translation will help many pupils around the world if the mathematics does not fit the culture. What is more, d'Ambrosio (1985) has indicated that 'recent advances in theories of cognition . . . show how strongly culture and cognition are related'. There is a growing body of knowledge about 'ethnomathematics', that is, mathematics in relation to social, economic and cultural background. The conventional view that mathematics is 'culture-free' is true in the sense of the universality of truth of mathematical ideas but it ignores the cultural basis and derivation of knowledge. It is thus, clearly, also important to look at the relationship between ethnomathematics and cognition if one is attempting to improve the mathematical competence of pupils around the world. Ethnomathematics is the mathematics practised among identifiable cultural or subcultural groups. It might be the only mathematics which makes sense within such groups. Perhaps this reinforces the view that curriculum change is necessary, and that it may not be sufficient simply to look at language considerations. Gay and Cole, d'Ambrosio and Gerdes (1988) all stress the importance of beginning with materials from the indigenous culture and utilizing them to extract the universal truths of mathematics. Which materials are used and in what ways will also have to take into account such issues as gender roles, if the mathematics is to be accepted by all pupils. Lancy (1983) has, in fact, gone so far as to propose an alternative stage theory for cognitive development to that of Piaget in which the sensori-motor and pre-operational stages are succeeded by a stage in which cognitive growth has 'much to do with culture and environment and less to do with genetics'. Bishop (1988) has suggested that it is at this stage that different cultures develop different mathematics.

It is only possible, in the space available, to provide a brief hint of the extent of problems of language and mathematical education, for a very considerable body of research has been documented (Wilson, 1981). However, many questions remain unanswered. Do bilingual children suffer academically when forced to learn in their weaker language? There is some evidence that mechanical arithmetic does not necessarily suffer but that, perhaps not surprisingly, arithmetic derived from word problems does. Word problems will be considered more fully in the next section. Is it possible that learning mathematical concepts in two languages could help to free the concepts from dependence on language thus enhancing understanding, or would it depend which two languages? How does one teach mathematics in a language which lacks essential mathematical vocabulary? How does one teach mathematics when essential ideas are not present? How does one take account of ethnomathematics? These questions and the problems expressed earlier are not completely irrelevant for teachers who imagine they speak the same language as their pupils. Studies of problems of language and mathematical education might ultimately enlighten us all in terms of problems experienced by children, in particular those whose language is much more restricted than our own.

WORD PROBLEMS

A word problem, or verbal problem, is simply a question which requires the application of mathematics in order to achieve a solution, but in which the required procedure has first to be extracted from within sentences. These sentences are usually intended to provide a real-life setting for the problem, but often teachers believe they confuse the pupils. Thus:

> Sarah had 5 pennies. Her father gave her another 3 pennies. How many pennies did Sarah have altogether?

is a simple word problem which requires the use of the elementary procedure

$$5 + 3 = 8$$

to achieve the solution. In Chapter 3, problem-solving was described as what goes on when a learner strives to find the solution to a novel problem—in fact, when previously learned knowledge, rules, techniques, skills and concepts have to be combined in a new way. Word problems are often not novel, being frequently, for the teacher, simply another way of providing practice of simple algorithms. Thus word problems and problem-solving need to be clearly distinguished. A problem which requires 'problem-solving' because of its novelty is likely to be described in words within sentences. A word problem might provide particular children with novel problem-solving opportunities. But usually, when the difficulties of word problems are being discussed, all that is meant is the kind of elementary practice problem given above. Thus most of the issues of problem-solving dealt with in Chapters 3 and 6 are not relevant in the present discussion. Many teachers, particularly from countries in which English is not the first language of the pupils, express great concern about the difficulties which their children experience when handling word problems. This is hardly surprising. Even the simple example above might confuse some pupils whose first language is English, never mind those pupils struggling with a second language.

One assumption of mathematics teachers has been that pupils first need to be confident in handling the purely numerical form before they are ready to attempt the equivalent word problem. Carpenter and Moser (1982) have, however, repudiated the suggestion that 'verbal problems are difficult for children of all ages, and [that] children must master addition and subtraction operations before they can solve even simple word problems'. Their research suggests that many children can solve basic word problems involving addition and subtraction before they receive any formal instruction at all. It is interesting to compare this with the findings reported by Hughes (1986) which confirm that, with small-sized sets, young children could perform addition and subtraction as long as real-life situations were being described, in other words the question was not 'disembedded', to use the terminology of Donaldson (1978). Word problems, of course, specifically set out to embed the arithmetic within a real-life context. In the research reported by Carpenter and Moser the strategies used by children in addition problems were based on counting (variations on counting all and counting on) and subtraction was based on separating and matching, together with counting techniques. The results indicated that children continued to use informal techniques based on counting well into the middle years of schooling, but that eventually most children began to use number facts and algorithms. The major problem which emerged was that

> by the age of 9, many children mechanically add, subtract, multiply, or divide whatever numbers are given in a problem with little regard for the problem's content. Somehow in learning formal arithmetic procedures, many children stop analysing the problems they attempt to solve.

It seems, therefore, that it is the transition from informal procedures which the child constructs to the procedures which the teacher teaches and expects the child to learn where the difficulties arise. This appears to be more a problem of learning than one of understanding the language of the question. It is certainly necessary to rethink the assumption that the numerical algorithms must come before the corresponding word problems. It is possible that the better way to learn is through word problems.

All research into performance on word problems has revealed the enormous variety of sentence and overall problem structure which is possible. Carpenter and Moser described 17 different kinds of elementary word problems involving only addition and subtraction, and these were still not completely unambiguous. A more complete framework by Vergnaud (1982) extends word problems to operations on integers. Perhaps because of this complexity, Laborde (1990) has attempted to list descriptively the main variables of word problems, and these were given as:

> how relations between the given and the unknown quantities are expressed, and in particular the degree to which they are made explicit;
>
> the order of items of information;

the degree of attraction of some words, such as the priority of numbers over words or the use of keywords like 'more', 'less' related to arithmetical operations, which may be distractors as well as cues;

the complexity of the syntax and of the vocabulary.

This analysis appears to be particularly valuable when looking into whether changes in wording will lead to improvement in performance. One of the 17 examples provided by Carpenter and Moser is:

> Connie had some marbles. She won 8 more marbles. Now she has 13 marbles. How many marbles did Connie have to start with?

Laborde quotes research by De Corte which makes clear how important is the first sentence, which could have been omitted in a more abbreviated form of the same question. The suggestion is that inexperienced pupils are more dependent on text-driven processing whereas more mature pupils have mastered complex semantic problem schemes. Thus, 'rewording verbal problems so that the semantic relations are made more explicit . . . facilitates the construction of a proper problem representation'. It is now clear that the order of information, the relations between known and unknown and the transition from known to unknown all influence understanding of a word problem in younger learners. Thus, it should be possible to improve performance on word problems by amending the wording.

All evidence suggests that it is not a simple matter to explain children's difficulties with word problems, so it is not easy to find ways of improving performance. Two ways have emerged, but neither is easy to implement. The more obvious way is to make the language more comprehensible, but all relevant research has indicated how complex and varied are the possible linguistic structures which might be employed in setting up a word problem. However, teachers should take note of the importance of thinking about wording and should note that there is evidence which will enable them to set problems in which the wording interferes less than it might otherwise. The other way of trying to improve performance on word problems is to attempt to build better on the informal methods employed by children before formal instruction has modified and possibly confused their thought processes. This is even more difficult to implement because little is known about when is the optimum moment to attempt to introduce any new procedure, and also about possible detrimental effects of allowing children to continue to use informal methods for as long as they wish.

SUGGESTIONS FOR FURTHER READING

Austin, J. L. and Howson, A. G. (1979) Language and mathematical education. *Educational Studies in Mathematics* **10**, 161–97.
Barnes, D. (1976) *From Communication to Curriculum*. Harmondsworth: Penguin Books.
Donaldson, M. (1978) *Children's Minds*. Glasgow: Fontana/Collins.
Harvey, R. *et al.* (1982) *Language Teaching and Learning 6: Mathematics*. London: Ward Lock.
Shuard, H. and Rothery, A. (eds) (1984) *Children Reading Mathematics*. London: John Murray.

Skemp, R. R. (ed.) (1982) Understanding the symbolism of mathematics, *Visible Language*, **XVI** (3).
UNESCO (1974), *Interactions Between Linguistics and Mathematical Education*. UNESCO/CEDO/ICMI.

QUESTIONS FOR DISCUSSION

1 How should we best act on the recommendation in the Cockcroft Report (*Mathematics Counts*) that mathematics teaching at all levels should provide opportunities for discussion?
2 Can mathematical concepts be learned without using language?
3 Discuss the suitability of a textbook you use from the points of view of vocabulary, readability, symbolism and the communication of meaning.
4 What lessons can we learn about the interference of language in mathematics-learning from the experiences of countries around the world?

Chapter 9

Is There a Theory of Mathematics-Learning?

MATHEMATICS AND THEORIES OF LEARNING

The place of theory in supporting and enlightening the process of learning mathematics has been discussed in Chapter 1. Debate about how mathematics is learned has continued throughout the recorded history of mathematics teaching, yet the process is still not founded on a universally accepted theory. Shulman (1970) claimed that 'mathematics instruction has been quite sensitive to shifts in psychological theories', but also that 'mathematics educators have shown themselves especially adept at taking hold of conveniently available psychological theories to buttress previously held instructional proclivities'. Some mathematics teachers and educationists have been very keen to look to learning theory for help in determining classroom practice, others have not been aware that there were theories, and yet others have reacted strongly against any suggestion that psychology could possibly have anything to offer. The concern of some teachers that there appears to be a variety of different theories and that it is difficult to know which is the correct one was part of the discussion in Chapter 1. The problem with a universally accepted theory would, of course, be that many teachers might then feel pressure to change teaching methods, and such pressure is not always a welcome outside influence.

In searching for appropriate theoretical underpinning, two kinds of theory demand attention. First, there might be theories which are specifically concerned with learning mathematics, and secondly there might be general learning theories which are applicable to learning mathematics. Given the complexity of the nature of human abilities and the fact that it is so difficult to isolate mathematical ability from other abilities and from overall ability (see Chapter 7), it would seem reasonable to assume that a general theory of learning might have as much to offer as any specific theory. General theories of learning certainly cannot be ignored. The theoretical approach to learning known as behaviourism has already been discussed in Chapter 4 and is an example of a general learning theory which led to the specific application to mathematics (see for example Thorndike, 1922). On the whole, behaviourism is out of favour, in fact it never has been popular with British educationists despite the widespread use of teaching methods which appear to be more closely related to behaviourist beliefs than to any other. Dienes (1973) certainly appeared to want to believe that 'no one today doubts any more the fact that the stimulus-response relation leads to a training which most of the time induces mental blockages'. As an alternative to behaviourism Bruner promoted discovery, managing to 'capture its spirit, provide it with a theoretical foundation, and disseminate it' (Shulman, 1970). Yet Novak (1977) felt obliged to state that, 'Unhappily, behaviourist doctrines continue to flourish in education.' Stewart (1985) endorsed the current view of most educationists in declaring that 'behaviourism was essentially finished as a theory that could adequately explain the more complex aspects of human mental activity'. Not surprisingly, therefore, modern attempts to develop theories of learning mathematics have adopted a cognitive psychology approach.

The work of Piaget was an important landmark in the development of cognitive learning theories, though he did not attempt to present his ideas as a learning theory. Lunzer (1976) discussed how far Piaget's results took us towards an epistemological theory of mathematics-learning with a view to the construction of a more adequate theory. Ausubel (1968) has presented a comprehensive theory of learning which demands consideration, incorporating results and concepts described by Piaget at the same time as criticizing the whole-hearted belief in the efficacy of discovery learning. An example of a specific theory of mathematics-learning is that by Dienes (1960), and it is necessary to consider how far this takes us towards a comprehensive theory. Contemporary developments embrace both the constructivist approach, building on the work of Piaget, Ausubel and Kelly (1955), and the information-processing view of cognitive development which pays attention to 'how the computer as metaphor affects our understanding of the processes of learning and teaching' (Kilpatrick, 1985). A somewhat independent approach to mathematics-learning has also been presented by Davis (1984).

A THEORY OF MATHEMATICS-LEARNING

The concept of place-value introduces difficulties for many children, and it is relevant to consider what is the most appropriate sequence of learning situations which might be utilized to promote learning. The two major alternative theoretical approaches which have been referred to in this book are the behaviourist and the cognitive. A behavioural approach suggests the use of stimulus-response situations through which connections are practised, but it is difficult to see how the underlying structure which is place-value could be grasped in this way, and much might depend on the quality of subsequent reflection by the child. A cognitive approach suggests that children should be placed in a learning environment in which they might investigate, and perhaps discover, and in which understanding might be constructed through their own efforts. The work of Piaget has been interpreted as suggesting that children learn only slowly and that they learn by abstracting from concrete situations in which they have been actively involved. The Multi-base Arithmetic Blocks (MAB) of Dienes thus provide a suitable early-learning environment enabling the construction of the understanding of place-value to take place.

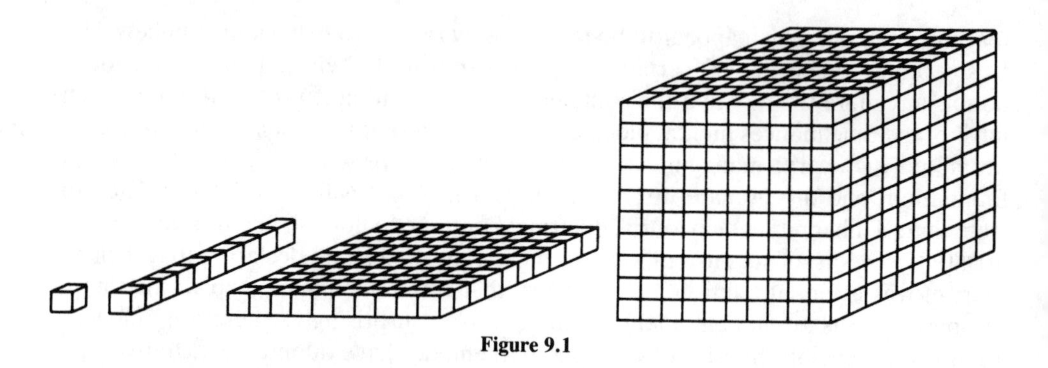

Figure 9.1

The MAB equipment consists of units, longs, flats and blocks in a wide variety of number bases. The base ten shapes are shown in Figure 9.1. If children are allowed to play with this equipment it might be considered reasonable to assume that they will eventually notice and appreciate (perhaps even without teacher intervention) that ten units are equivalent to a long, that ten longs are equivalent to a flat and ten flats are equivalent to a block. The structure of the apparatus will become apparent. Furthermore, children might begin to appreciate the foundations of simple arithmetical calculation through activities which involve exchanging shapes. Thus, for example, 13 units could become one long and 3 units, and the paying of a forfeit of 4 units from a long would leave 6 units. Teachers of young children will know that teaching based on such activities can, indeed, take a very long time. In fact, more equipment might be needed, for the structure is not dependent on the particular materials used, so matchsticks, bundles of ten matchsticks, and boxes containing ten bundles might provide a parallel activity. Other, similar, activities can easily be devised to promote the understanding of place-value in base ten. Some teachers would be concerned to divorce the place-value concept from the base ten concept by providing activities which depend on other bases, like six 'eggs' to an egg box. Dienes, in fact, provided MAB equipment in a wide variety of number bases for just this same reason. This outline of the teaching of place-value illustrates the practical application of the theory of mathematics-learning proposed by Dienes (1960).

Dienes began from the premise that mathematics could not be learned in a stimulus-response way because it was not mathematical content that provided the problem, it was the fact that mathematics-learning was so bound up with understanding structure. Although the equipment suggested by Dienes is comparatively well known to teachers (see also Seaborne, 1975) it is not so widely appreciated that the apparatus was proposed by Dienes, at least in part, as a way of putting the Dienes Theory of Mathematics-Learning into practice. In addition to the MAB, Dienes commended the use of the Algebraic Experience Material (AEM), the Equalizer (Dienes' Balance) and the Logical Blocks, all of which enabled mathematics-learning to be achieved as a constructive activity. Dienes drew his initial inspiration from the work of Piaget, Bruner and Bartlett, but his theory was also based on research which he himself had carried out. The resulting theory of mathematics-learning consisted of four principles:

1. the dynamic principle
2. the constructivity principle

3. the mathematical variability principle
4. the perceptual variability principle.

Amongst many other conclusions, the work of Piaget has been taken to suggest that children learn much more slowly than we had ever before imagined and that time needs to be allowed for concepts to form. In addition, learning is an active process, so concept formation is promoted by providing suitable learning environments with which children can interact. The dynamic principle of Dienes was a derivation from these Piagetian principles. Dienes, in fact, referred to 'Piaget's three stages in the formation of a concept', and described these as the play stage, the structure stage and the practice stage. The play stage was basically to be thought of as unstructured activity, so in terms of place-value it was playing with the MAB equipment. Eventually, as the realization of structure began to grow, children's activities could be much more structured and teacher intervention could ensure that the structure was grasped. Practice of the structure could then ultimately lead to more overt use of practice exercises, leading to simple arithmetic and the written recording of calculations using place-value. The play–structure–practice sequence was, of course, to be seen only in relation to a single concept. The practice activities for one concept might then become the play activities for another concept. It should also be pointed out that, for Dienes, the three stages subsequently became six (Dienes, 1973), and also that the play stage might not always seem, to older pupils, to be play.

It is appropriate, in the context of three stages in learning mathematical concepts as originally proposed by Dienes, to draw attention to the three stages described by Bruner. In terms of representing the world, or translating experience into a model of the world, Bruner (1966) suggested that the stages were the 'enactive', the 'iconic' and the 'symbolic'. These stages were further elaborated in Bruner *et al.* (1966). Many forms of knowledge can only be learned in an active way, like riding a bicycle or playing a sport. The early phase of learning an abstract concept, like place-value, might also require an enactive approach, with children engaged in manipulating concrete apparatus. A second approach to learning is, however, the use of pictures or visual images of some kind, so eventually, when learning about place-value, the actual concrete objects might be replaced by pictures of the objects. Textbooks, workcards and other written materials are heavily dependent on an iconic approach. Certain mathematical ideas might be learned directly from pictures and without any prior dependence on enactive representation. The ultimate approach to learning is symbolic, through language and through other symbols of a specific mathematical nature. Some mathematical concepts might be learned directly through the manipulation of symbols and without prior dependence on either enactive or iconic approaches. The three stages may be regarded as a sequential approach to learning, as perhaps in the case of place-value, with concrete equipment followed by pictures, followed by pencil and paper sums. Alternatively, the three forms of representation might be regarded as three different approaches to learning, with their appropriateness related to characteristics of the particular learner, such as prior experience and knowledge. Bruner did not suggest that there was any direct connection between the enactive, iconic and symbolic stages of learning new concepts and the stages of intellectual development suggested by Piaget. Certainly, any feeling that the appropriateness of a particular form of representation is related to the age of the learner would indicate a developmental aspect to the theory which does not appear to have been intended by Bruner. Nor have Bruner and Dienes indicated that they saw any relationship between their respective stages. With Piagetian theory also to be taken into account it seems there are almost too many stages for anyone to contend with and of which to make sense. Nevertheless, the stages of Bruner, as sequential forms of representation, and the stages of Dienes, as one of a number of teaching recommendations, are both, independently, valuable ideas. In their different ways they both draw to the attention of the teacher the need to cater for individual pupils in particular ways.

Dienes used the work of Bruner and Bartlett, together with his own research, to support the view that mathematics is a constructive activity for children and not an analytic activity. Formal logical thinking, dependent on analysis, may well be a pursuit in which adults can engage, but children need to construct their knowledge. In the case of the place-value concept, this is done by using a variety of forms of concrete activity in a variety of different number bases. It is interesting to speculate as to when we become capable of analytic thinking. Is there a relationship between Piaget's concrete operational stage and the need for children to construct their own understanding, and likewise between the formal operational stage and the ability to think analytically? Dienes did not refer to Piagetian stage theory as such in his exposition of the theory of mathematics-learning.

The issue of how to accelerate mathematics-learning was answered by Dienes in terms of providing varied learning experiences. Earlier discussion of concepts in this book (Chapter 3) has drawn attention to the fact that 'concepts describe some regularity or relationship within a group of facts' (Novak, 1977), and that concepts are also learned from examples and counter-examples (Skemp, 1971). Dienes cited Castelnuovo and Wertheimer in stating that, 'A mathematical concept usually contains a certain number of variables and it is the constancy of the relationship between these, while the variables themselves vary, that constitutes the mathematical concept.' This led Dienes to the mathematical variability principle. In terms of place-value, it was important to Dienes that children should work with a wide variety of number bases. In terms of learning about parallelograms, another example given by Dienes, it was essential that lengths, angles and orientations should all be varied. In fact, orientation has often not been varied in the experience of many children, and the belief that a square in certain orientations was not a square but was a 'diamond' was mentioned in Chapter 3. Teachers of less able children are often unconvinced by the suggestion that a variety of number bases is essential, believing that such an approach confuses the children. The exhortation to apply the mathematical variability principle in teaching about geometrical shapes, however, cannot be ignored.

Another issue considered by Dienes was that of individual differences, the subject of Chapter 7 of this book. This led to two recommendations, one being to organize learning on an individual or small group basis, perhaps using workcards, and the other being the perceptual variability principle. It was considered important that the perceptual representation of a concept should be varied, thus, in learning place-value, the specifically provided wooden blocks of the MAB would not be sufficient. The idea of matches or straws and bundles of matches or straws has already been mentioned. Some writers have suggested different coloured counters (5 yellow = 1 green, 5 green = 1 red, etc.) but there are disadvantages with this equipment because, perceptually, one counter does not look equivalent to five of another colour. Dienes also suggested that some of the AEM material, based on shapes other than cuboids, was appropriate for place-value. In the case of parallelograms, these may be represented on paper and blackboard, may be made out of wood, metal and plastic, may be outlined with pegs on a pegboard or elastic bands on a nailboard, be made by putting two congruent triangles together or by dissecting a rectangle, and be seen in shapes in the real world—in windows and other features of buildings and floors and in patterns on wallpaper, and many other designs. The need for 'variability', in both mathematics and materials, is often referred to as the principle of 'multiple embodiment'.

The four principles of the Dienes' Theory of Mathematics-Learning were not intended to apply only to concepts of elementary mathematics. One of the most difficult aspects of slightly more advanced and abstract mathematics is algebra. Attention has already been drawn to the use of certain Dienes AEM wooden blocks to promote an early understanding of $(x + a)^2 = x^2 + 2ax + a^2$ (see Chapter 5). Whether earlier understanding than normal is sought or not, there can be no doubt that equipment can be used to approach quadratic expansions constructively rather than abstractly. The constructive approach might involve placing patterns of pegs in a pegboard, as in Figure 9.2.

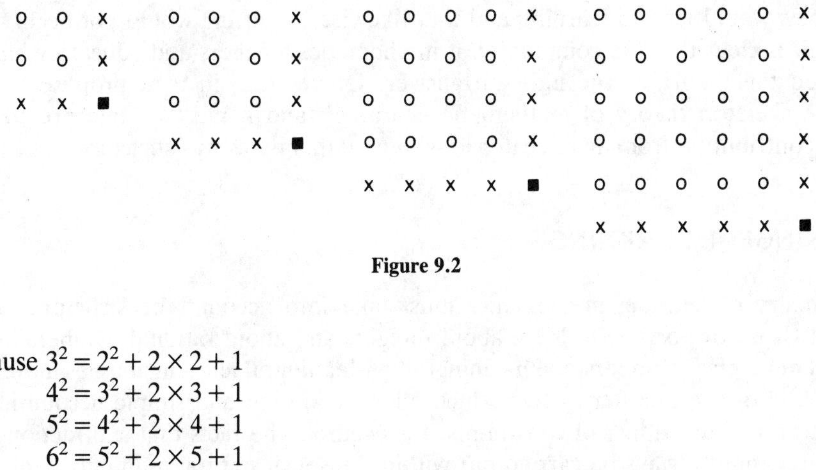

```
o o x    o o o x    o o o o x    o o o o o x
o o x    o o o x    o o o o x    o o o o o x
x x ■    o o o x    o o o o x    o o o o o x
         x x x ■    o o o o x    o o o o o x
                    x x x x ■    o o o o o x
                                 x x x x x ■
```

Figure 9.2

Because $3^2 = 2^2 + 2 \times 2 + 1$
$4^2 = 3^2 + 2 \times 3 + 1$
$5^2 = 4^2 + 2 \times 4 + 1$
and $6^2 = 5^2 + 2 \times 5 + 1$

we might deduce that, in general,

$$(x + 1)^2 = x^2 + 2x + 1.$$

The formula has been constructed, not proved. The numbers have been varied, but the structure remains the same, so mathematical variability has been applied. The same result can be approached using square regions on a nailboard, coloured regions on ordinary squared paper, or Dienes MAB equipment, so perceptual variability can be applied. The result can be extended, by a process of construction, to

$$(x + a)^2 = x^2 + 2ax + a^2,$$

and to

$$(ax + 1)^2 = a^2x^2 + 2ax + 1,$$

and so on, until a whole range of quadratic expansions has been explored, and appropriate generalizations constructed.

The Dienes Theory of Mathematics-Learning is very satisfying in a number of ways. It clearly slots into the cognitive approach to theories of learning, building on the work of Piaget and Bruner, so it has a base in one of the two major areas of learning theory—the one which currently receives more support from educationists. Certain other important issues like how to accelerate learning and how to cope with individual differences are incorporated. Current views on learning are placing considerable emphasis on the belief that knowledge is constructed by each individual and often cannot be simply transferred ready-made from teacher to learner. The constructivity principle of Dienes can perhaps be seen as a primitive view on constructivism. But it must also be admitted that the theory has limitations. The constructivity principle relates to learning individual concepts, and the relationship between the learning of a new concept and the existing knowledge structure already held in the mind is not considered. Mathematics is, after all, a very hierarchical subject in which new knowledge generally must be linked on to existing knowledge; if prerequisites have not been mastered the new knowledge just cannot be learned. Nor was the issue of readiness explicitly tackled by Dienes; it was tacitly assumed that adopting the four principles would lead to learning and that likewise forgetting would not occur. Certainly, it is clear that the community of mathematics teachers and educators has not accepted the theory as the ultimate answer. Dienes did, in fact, propose it as 'a feasible skeleton theory of mathematics-learning', and it must be interpreted as a useful contribution from an educator to whom mathematics was structure.

MEANINGFUL LEARNING

Any theory of learning mathematics must take into account the structure of the subject. It is not possible to learn about integers and about rational numbers before natural numbers are understood meaningfully. Meaningful learning implies more than knowledge of the number system which allows counting and simple accounting. It implies an understanding of constraints, for example the facts that subtractions and divisions cannot always be carried out within the set of natural numbers. When the existing knowledge structure is sufficiently rich and varied, and better still when the child is asking questions which require new input of concepts, the time is right for injecting these new ideas. If an attempt is made to force children to assimilate new ideas that cannot be related to knowledge which is already in the knowledge structure the ideas can only be learned by rote. A range of examples from mathematics will illustrate this point.

The algorithm for calculating the arithmetic mean is one of the simplest of all algorithms in mathematics. It is so simple that it is all too easy to teach without paying due attention to linking the algorithm in a meaningful way to existing knowledge. Without such links the algorithm will be learned by rote, will be easily forgotten, and will not promote flexibility of thinking. Basically, the arithmetic mean is one of a number of measures associated with the idea of a representative value. If a class were to write to a class in another country and regularly exchange information and news they might wish to include some statistical information. There would be a number of ways of sending information about height, along the lines of 'we are about 150 cm in height', or 'a lot of us are exactly 148 cm in height (to the nearest cm)'. The ideas of mean, median and mode must be seen as attempts to convey information about a population and the comparative value of these three statistics would need discussion. Another situation familiar to children concerns sharing out sweets. If, at Christmas, the teacher were to leave a tin of sweets for the children to help themselves, different children would take different numbers of sweets, and that would not be considered fair. Instead of Stephen having eight, Trudy having seven, Helen having three and Craig having two they should all really have the same number of sweets, which we could calculate by putting all the sweets together and sharing them out. The abstract equivalent is to add up all the numbers of sweets in the possession of all the children and divide by the number of children. It is possible to link the idea of arithmetic mean to previously held knowledge, thus conveying ideas in a meaningful way.

The introduction of the sine and cosine ratios to secondary school pupils needs to be linked to a number of ideas, including similarity (or enlargement) and ratio, triangles, right angles, other angles and lengths. The real meaning and purpose of sines and cosines will not be absorbed if these two ratios are not linked with what should have been learned earlier and with some kind of motivating force such as the need to be able to perform certain kinds of calculation from certain sorts of data. Often either the motivation or the previous knowledge is not there. In terms of prior knowledge, both similarity and ratio are difficult ideas, and may not be adequately formed. One must, however, acknowledge that they might become better formed through a study of elementary trigonometry, but if there is no relevant knowledge there at all, sines and cosines would once again have to be learned by rote. The problem of motivation is not easily solved either, for different real-life situations may be meaningful to different pupils. A link with prior knowledge, however, certainly needs to be sought.

Certain mathematical knowledge is so basic that there is unlikely to be any relevant knowledge already in the mind to which new ideas can be linked. Very young children usually enjoy trying to slot odd-shaped wooden or plastic objects through holes in the top of a box, designed in such a way that only one shape will fit into any one hole, and then only with one particular orientation. It is a process of discovery for children to solve this problem, but eventually they become quite proficient and their interest wanes. This particular game has, however, taught them a great deal of basic spatial knowledge. At a later stage in life pupils might be given Cuisenaire (or other) rods to play with, and will discover that red + pale green = yellow, and orange − pink = dark green (without the symbolism!). Without such a period of discovery using coloured rods, or with beads or counters, or with equipment of some other kind, it is difficult to see how children could learn the basis of number bonds. The only alternative would appear to be rote. Certain mathematical knowledge is so basic that there might not be any part of the existing knowledge structure with which it could be connected.

The theory of meaningful learning proposed by Ausubel (1968) was a general theory and was not specific to mathematics. It incorporated the ideas presented above and so, to Ausubel, meaningful learning was a process through which new knowledge was absorbed by connecting it to some existing relevant aspect of the individual's knowledge structure. If there were no relevant concepts already in the mind to which new knowledge could be linked, the new knowledge would have to be learned by rote and stored in an arbitrary and disconnected manner. If new knowledge was assimilated within the existing knowledge structure as a related unit, and if appropriate modification of prior knowledge (accommodation) took place, the result was meaningful learning. It was therefore not necessary for all, or perhaps even much, knowledge to be acquired by a process of discovery. Good expository teaching could ensure that new knowledge was linked to relevant existing ideas, and this might not only be more economical (in terms of time taken) than was discovery, it might be more efficient in terms of quality and breadth of learning. If you really could ascertain what the learner already knew, you would then know what and how to teach. Discovery learning would be necessary with very young children, and at this stage of life the emphasis would need to be on encouraging concept formation rather than teaching for concept acquisition. But once a rich structure of knowledge has been learned the most efficient way to proceed would be by exposition. Discovery methods might occasionally be appropriate, but meaningful verbal learning could, in most circumstances, be at least as effective and in some ways better than any other method.

Clearly this theoretical stance can only become reality if one can find out in sufficient detail, and in a reliable way, what the learner already knows, and if one can then ensure good, as opposed to indifferent or bad, expository teaching. Given good teaching, if subject matter was inadequately learned the reason would then be that pupils did not have the required foundation of relevant knowledge on which to anchor new ideas. It would, of course, also be necessary for *all* members of a class to have the required foundation of knowledge, something which is not easy to obtain in the normal school situation. Such difficulties would not invalidate the theory, but would raise serious problems for the teacher.

Ausubel's theory of meaningful learning contained a number of other ideas which will require discussion in due course, but first the relationship between the ideas of Ausubel and Piaget demands attention. Ausubel used data collected by Piaget, accepted the ideas of assimilation and accommodation, and from time to time referred to 'concrete' and 'formal' or 'abstract' stages, without accepting the full implications of Piagetian stage theory. Novak (1977), whose own work ably explained and clarified Ausubelian theory, claimed 'no operational conflict exists between the ideas of Piaget and Ausubel'. In terms of readiness, Ausubel's view was closer to that of Gagné than to that of Piaget. The existing parts of the knowledge structure to which new learning needed to be linked were called 'subsumers' or 'subsuming concepts' by Ausubel; subsequently they became known also as 'anchoring' ideas or concepts. So, if the subsumers or anchoring ideas were there, the pupil was effectively ready. Readiness was only related to stage of development in its most open interpretation as dependent on having more and better developed subsumers. Shulman (1970) certainly expressed the view that Ausubel was in fundamental agreement with Gagné in that the key to readiness was prerequisite knowledge. Novak (1977), however, indicated that he thought Ausubel's view on readiness was close to that of Bruner. Perhaps this can be taken as an indication of the reconciling power of Ausubelian theory! To Ausubel, even if the child was not ready in the sense of having appropriate subsumers, all was not lost. There was then the possibility of using an 'advance organizer' to bridge the gap.

Matrix multiplication can appear very arbitrary, complex and meaningless to pupils, and therein lies a recipe for disaster in terms of meaningful learning. Despite attempts to motivate the idea through using shopping bills and the like, despite attempts to base the introduction on transformation geometry, there are arbitrary aspects to the procedure. Matrix multiplication is, however, essential to the long-term development of an understanding of modern algebra, and it may be applied, as a technique, in a number of different topics in school mathematics. Different authors have used a variety of different ways of introducing matrix multiplication, but Matthews (1964) used a very ingenious method. Secret messages were to be coded for transmission by: (a) representing each letter by a number; (b) changing the messages to strings of numbers; (c) grouping the numbers in fours to make 2×2 matrices; and (d) applying an encoding 2×2 matrix to each matrix of the message to convert the original string of numbers into another string which bore no resemblance to the message. The messages were then despatched as strings of numbers which could not be decoded without applying the decoding matrix which was, of course, the inverse of the encoding matrix. For many children the whole activity was good fun. The object of the whole exercise was, of course, not to teach how to send coded messages, but was to persuade children to master an arbitrary rule. Ultimately, this rule would be seen to be needed in more mainstream curriculum mathematics, but pupils might not be so motivated to learn about matrix multiplication in contexts such as simultaneous equations and transformation geometry without being able to see where it was all going. Having implanted this arbitrary and disconnected knowledge in the mind of the learner there was then an anchoring concept onto which to latch more important applications of matrix multiplication. In a sense, the use of matrices to send and decode messages was an advance organizer.

To Ausubel (1960), advance organizers were 'more general, more abstract, and more inclusive' than the ideas and knowledge which were to follow. It is therefore doubtful whether sending coded messages would satisfy strict Ausubelian criteria for an advance organizer for matrix multiplication. The use of advance organizers defined much less rigorously is probably quite a common teaching technique, but finding advance organizers which satisfy the more rigorous criteria of being more general, more abstract and more inclusive is not so easy. Scandura and Wells (1967) translated the idea of an advance organizer into 'a general, non-technical overview or outline in which the non-essentials of the to-be-learned material are ignored'. The idea of an advance organizer is certainly too useful to be rejected for technical reasons, so perhaps any idea which we can put into the minds of learners which will act as a bridge for subsequent, more detailed knowledge should be regarded as an advance organizer. Novak (1977) claimed that 'research studies that focus on the use of various forms of advance organizers . . . are not profitable'. The hierarchical nature of mathematics would also appear to suggest that there should not be many occasions when new knowledge cannot be linked to existing knowledge, but the idea of the advance organizer is still a valuable one to keep in mind.

Concept maps were introduced in Chapter 3. The psychological justification for using concept maps can now be seen in relation to meaningful learning and the relating of new knowledge to an existing knowledge structure. Ausubelian theory must be regarded as an original source for the idea of concept maps, though it has been Novak (1980), Novak and Gowin (1984) and many others, who have advocated their use in recent years. To Novak and Gowin a concept map is 'a schematic device for representing a set of concept meanings embedded in a framework of propositions . . . [which] work to make clear to both students and teachers the . . . key ideas they must focus on for any specific learning task'. When the learning sequence is completed they 'provide a schematic summary of what has been learned'.

SUPERORDINATE AND SUBORDINATE LEARNING

The organization of knowledge in the mind demands constant review and rearrangement, 'a pushing and pulling of concepts, putting them together and separating them' (Novak and Gowin, 1984). It involves the realization that a particular conceptual structure may be differentiated into concepts which might, in one sense, be considered subordinate. It involves the realization that certain ideas are all part of a more inclusive or superordinate concept structure. Skemp (1971) discussed the ideas of primary concepts which were 'derived from our sensory and motor experiences of the outside world', and secondary concepts which were 'abstracted from other concepts'. Certain concepts were seen to be of 'higher order' than others, which implied they were 'abstracted from' others. Ausubel (1968) wrote of 'progressive differentiation' in learning, in which the most inclusive elements of a concept are introduced first and then the concept is dissected or progressively differentiated in terms of detail and specificity. He also wrote of superordinate learning, when previously learned concepts are seen to be elements of a larger, more inclusive, concept structure. The kind of reorganization of knowledge involved in learning mathematics is certainly likely to involve the two-way process of relating concepts both to subordinate and to superordinate concepts.

Early learning experiences in mathematics are largely concerned with developing competence and understanding in numbers and the 'four rules'. Considerable time is normally spent on the operations of addition, subtraction, multiplication and division. Over the years, these same operations are applied to fractions and to decimals and eventually their application to all real numbers should be mastered. Some students proceed to apply the same four operations to complex numbers, and a wider perception of, for example, multiplication is achieved. Most children study sets, and operations are introduced here, too, with union and intersection being the most likely, though not the only ones possible. Vectors are learned by many pupils, and the operations of addition, subtraction, and for some pupils scalar product, and perhaps even vector product, are taught. Operations are applied to matrices, through addition, subtraction and multiplication. Some students learn propositional calculus and use operations such as conjunction and disjunction. Eventually, and perhaps at some stage within the learning sequence above, the concept of 'binary operation' might be introduced. The only sensible way to approach the idea of binary operation with pupils is to have many examples of such operations on which to define the more inclusive concept (cf. the idea of multiple embodiments from Dienes). In this sense, the idea of 'binary operation' might be considered to be superordinate to 'multiplication'. In just the same way the concept of 'commutativity' would make little sense without examples of commutative operations and non-commutative operations (defined on particular sets) on which to build the more abstract idea. Superordinate learning appears to be very much a part of learning mathematics.

In contrast, the important concept of 'symmetry' is usually studied rather differently. On the basis of a few examples like human and animal faces, butterflies, inkblots, mirror reflections, and the like, the idea of symmetry as a kind of repeated regularity is introduced. Having introduced what is, in essence, bilateral symmetry in two dimensions, the possibility of other repeated regularities in familiar objects and in mathematical entities is researched. In some shapes the regularity is seen to be a rotational one, which leads to a differentiation into bilateral and rotational symmetry. Rotational symmetry itself, when analysed, leads to the idea of order. Both bilateral and rotational symmetry involve the idea of axes of symmetry. Further differentiation between two-dimensional and three-dimensional objects introduces another idea, that of planes of symmetry. Having developed the idea of symmetry by progressively differentiating a general idea of regularity, the learner is then able to look at symmetry in mathematics, and perhaps even in the natural world and the man-made world, with much greater insight.

Ausubel expressed the view that concept development proceeds best when the most general, most inclusive elements of a concept are introduced first and then the concept is progressively differentiated in terms of detail and specificity. A much quoted example is that, to young children, four-legged animals are all 'dogs', and it takes progressive differentiation to sort out which are cats, cows, horses, sheep and so on. The same is variously true about fish, about ducks in the park duck-pond, and about cars. Yet either there appear to be exceptions or it is the case that learning works both ways, from the superordinate to the subordinate and vice versa. Children learn what are apples, what are oranges, what are bananas and only subsequently come to know them collectively as fruit. In mathematics, they learn about squares and rectangles (and perhaps parallelograms, kites and even rhombuses) before learning about quadrilaterals. It is, in essence, a matter of whether all four-sided shapes are seen by the child as squares, the idea then being progressively differentiated, or whether squares and rectangles are seen to be different and quadrilaterals are perceived as a superordinate idea.

In fact, learning mathematics must involve both progressive differentiation and superordinate learning working together; treating the two ideas separately is merely a convenience to enable analysis of their meaning. The various different number sets and the relationships between them have already been used as illustrations a number of times in this book. It could be legitimately considered that learning about numbers involves progressive differentiation, but, equally, it could be said that a variety of different kinds of numbers is introduced over a period of time until, eventually, the superordinate concept of real numbers is introduced. Although learning about quadrilaterals appears to be an illustration of superordinate learning, the study of triangles appears to take place by progressive differentiation. At a higher level, the factorization of quadratic expressions has usually been tackled systematically, by gradually introducing more and more complicated collections of coefficients, and this appears to be progressive differentiation. Readers will probably have their own views on illustrations of progressive differentiation and superordinate learning. Novak (1977) stated that 'Determination of what in a body of knowledge are the most general, most inclusive concepts and what are subordinate concepts is not easy', so complete agreement about progressive differentiation and superordinate learning is unlikely. It is important, however, to consider relationships between concepts for, 'One reason school instruction has been so ineffective is that curriculum planners rarely sort out the concepts they hope to teach and even more rarely do they try to search for possible hierarchical relationships among these concepts' (Novak, 1977). Clearly, concept maps could play a part in curriculum planning which attempted to analyse the relationships between concepts.

CONFLICTS AND FAILURES IN LEARNING

There are times when conflict occurs in learning, and also when learning either does not take place or is quickly forgotten. All of these issues require consideration, and Ausubel has again provided us with a theoretical model. Conflict of meaning, termed 'cognitive dissonance' by Ausubel, might occur for many reasons. It might arise when our use of the word 'vertical' suggests a meaning which is in conflict with the previously understood idea. It might arise when one teacher implies that a triangle is a polygon and the textbook claims it is not. It might arise when one mathematical text provides a definition of natural numbers which includes zero and another book excludes zero. It might arise when the mathematical definition of gradient is seen to be different from the meaning of the concept in the real world. In any one of very many ways cognitive dissonance can occur. Essentially, this is a problem of accommodation, though rather different from most accommodation problems. Conflicting ideas create disequilibrium and somehow they must be reconciled, and this is achieved by the process of 'integrative reconciliation'. Without integrative reconciliation it is possible that learners might compartmentalize the conflicting ideas thus, for example, accepting that force and acceleration are in proportion in mathematics lessons but acting as if they think otherwise outside the school environment. In the case of gradient it is *necessary* to compartmentalize, with the attendant danger of legitimizing the holding of two different definitions for the same entity. It is not easy to prescribe for integrative reconciliation, but cognitive dissonance is a feature of school learning which must be acknowledged. It is especially difficult to achieve reconciliation when the cause of the conflict crosses subject boundaries, as in the case of there being one definition for histogram in mathematics and perhaps a rather different one in biology.

Reasons why learning does not take place include the non-cognitive, such as not paying attention at the critical time, and the cognitive, like not being ready in the sense of having adequate subsumers. The issue of forgetting is equally complex. In the first place there appear to be degrees of forgetting, for it is possible to forget but then recall everything when appropriate cues are presented, and it is also possible to forget apparently irretrievably. Novak (1977) stated that 'Most information we learn cannot be recalled at some time in the future', suggesting that forgetting is the norm and that it is remembering that requires explanation. Ausubel's theory explained variation in rates of forgetting in terms of the degree of meaningfulness of the learned material. In the case of material learned by rote, the expectation would be that it would be forgotten, probably sooner rather than later, because such knowledge must be stored in a part of the knowledge base which is unconnected to major structures. The learning of the vocabulary of a foreign language is almost inevitably at least partly by rote, but words are remembered better under certain conditions such as regular use in sentences (which introduces a degree of meaningfulness). Ausubel described 'over-learning', meaning repetition, revision, and perhaps some extension, and in this way rote-learned material might be retained for considerably longer than without over-learning. When knowledge has been acquired meaningfully the expectation would be that retention would be for very much longer. Forgetting can, however, still occur because of 'obliterative subsumption'.

When a new idea is introduced and becomes connected to relevant subsumers accommodation might lead to changes in the way both the new idea and the subsumers are understood. Subsequent new knowledge might then also produce change, both in the previous new knowledge and in the subsumers. This process continues throughout life so that, with successive modification and amendment, a body of knowledge or a conceptual structure might become so modified that it cannot be brought back to mind in its original form. This is a very neat theory which it is difficult to confirm or reject, after all, its verification requires examples of knowledge which has been forgotten! It is certainly possible that we have learned techniques and methods of solution which were valuable when introduced but which have been forgotten because subsequent techniques have, in subsuming them, effectively obliterated them. Most pupils who learn about quadratic factorization are provided with tips, rules, or processes to help with the difficult early stages. Eventually, the elementary techniques fall into disuse and can easily be forgotten altogether once greater experience and expertise creates a state in which factorization is no longer found to be difficult. One factorization technique is as follows:

Given $10x^2 + 23x + 12$
we require two numbers such that their product is 10×12 and their sum is 23, that is,

$$(\quad) \times (\quad) = 120$$
and $$(\quad) + (\quad) = 23.$$

The two numbers, found by trial and error, are 15 and 8, then:

$$\begin{aligned} & 10x^2 + 23x + 12 \\ &= 10x^2 + 15x + 8x + 12 \\ &= 5x(2x+3) + 4(2x+3) \\ &= (5x+4)(2x+3). \end{aligned}$$

Eventually, it becomes possible for many pupils to factorize entirely by inspection and the above method falls into disuse and may ultimately be forgotten. This illustration may not be ideal as an example of obliterative subsumption, but it does show how valuable knowledge may be obliterated without the learner becoming deprived.

The phenomenon of obliterative subsumption appears to suggest that meaningfully learned material often cannot be recalled in the exact form in which it was originally stored, whereas rote-learned material can only be recalled in the precise form as the original, since it cannot be subjected to obliterative subsumption. If that is true it indicates one advantage of rote-learned material, but all other advantages appear to be in favour of meaningful learning. Knowledge which is acquired in a meaningful way is retained longer than if acquired by rote, and it contributes to the growth and development of subsumers and therefore facilitates further meaningful learning. Teaching experience suggests that rote-learned material is *not* always recalled, by children, in the form in which it was learned, but this can be explained as forgetting (see Novak, 1977).

The learning theory proposed by Ausubel is extremely comprehensive and space has allowed the consideration of only a selection of the issues discussed by him (1968).

For the most part, mathematics educators have not paid much attention to Ausubelian theory, so the relationship to learning mathematics has not been sufficiently widely debated and few authors have provided a variety of mathematical examples in connection with the theory. Contemporary views on learning frequently draw from the work of Ausubel, as well as Piaget and Bruner, so it is helpful to know something of the views of all three if one is to be able to place current views in context. Science educators have paid rather more attention to Ausubel, but frequently their interpretations have filtered through to the classroom level only in terms of practical suggestions like the commendation of the use of concept maps. The very comprehensiveness of the theory of meaningful learning proposed by Ausubel suggests that we do now have a useful model with which any future learning theory might be compared in order to help assess its value. There have, of course, been critics of Ausubelian theory, for example some mathematics educators would react strongly against any suggestion that verbal or expository learning is as effective and as efficient as Ausubel claimed. This is difficult to determine anyway, because the key to the theory is that one must first ascertain what the learner already knows and then apply not only appropriate, but good quality, expository teaching. In the first place, it is difficult to ascertain in real detail what a learner already knows, and in the second place, if we are to accept recent reports on mathematics teaching, it is difficult to find examples of consistently good quality expository teaching. It is certainly the case that the strongest supporters of Ausubelian theory have always accused critics of not studying the theory in sufficient detail.

CONSTRUCTIVISM

Hughes (1986) has provided evidence that pre-school children are able to invent their own symbols and symbol systems to represent quantities, that is numbers of objects. Although it was admitted that some teacher interaction is necessary at times, nevertheless the evidence of the ability of children to invent appropriate notation, often based on one-to-one correspondence, is convincing. A suitable symbol for zero, that supposedly very difficult early concept, can also be invented, it was claimed. Yet these same children can experience great difficulty in coming to terms with conventional symbolism. Hughes has also drawn attention to the similarities between many of the children's own systems and number systems used by other cultures. Formal arithmetic presents children with symbol systems and methods of manipulation which are the products of hundreds of years of development and refinement. Why is it that children are capable of devising their own symbols, but find difficulty in coming to terms with real understanding, with the systems and processes which teachers try to impose?

Carraher (1985) has shown that Brazilian children with little or no formal education can invent their own methods of carrying out calculations in order to earn a living in the 'informal sector' of the economy. These mental calculations make sense to the child, being based on real transactions in which goods or services are sold for cash. Results appear to indicate that problems which do make sense in this way are more easily solved than the decontextualized ones of formal arithmetic. There is, of course, a difference in that the solutions to real-life problems involve the manipulation of quantities whereas school problems might appear to involve the manipulation of symbols without meaning. Nevertheless, some important questions are raised. School mathematics often assumes that children first need to learn the essential mathematical procedures before these procedures can be applied to verbal and real-life problems. Is this completely the wrong way round? If, as evidence seems to suggest, children are able to develop their own computational routines, why should we attempt to impose on them the more sophisticated conventional systems which appear not to be understood? Schliemann (1984) has also compared the problem-solving capabilities of professional carpenters and their apprentices. The unschooled professionals sought realistic solutions to real problems and were comparatively successful. The schooled apprentices seemed to be inclined to treat the problems as school assignments and were often wrong; what is more, they were unable to appreciate when they had produced an absurd answer.

These examples have been selected to draw attention to the dilemma that children often seem capable of constructing mathematical knowledge for themselves which is meaningful and helpful in the real world when school-taught knowledge might be misunderstood, misapplied and even rejected. Perhaps we do not take sufficient account of the nature of learning. A considerable amount, some might say the majority, of the teaching that takes place in mathematics classrooms seems to be based on the view that it is possible to transmit knowledge from teacher to learner, and that what is received is an exact copy of what was transmitted. Yet teachers know that this is not the case. A major reason why children fail to achieve lasting learning is that the knowledge was never comprehensively grasped in the first place. Transmission learning often only achieves limited success, and the severity of the limitations may not be discovered until much later, or may even never be discovered. Each individual child is likely only to receive a subset, possibly a unique subset, and probably quite a small subset, of what was transmitted. The alternative view that, if placed in a suitably rich environment, children will discover mathematics for themselves has been considered earlier. Often the outcomes from discovery learning environments are almost as disappointing as from transmission teaching, the main benefit seeming to be that the children have enjoyed the change! Guided discovery, in particular, has many advocates but input from teachers is implied, it being in a way merely a combination of discovery and transmission. The problem is that we cannot ignore 'one of the fundamental assumptions of cognitive learning psychology [which] is that new knowledge is in large part constructed by the learner' (Resnick and Ford, 1984). This assumption is the basis of constructivism. In the last resort, knowledge has to be constructed (or reconstructed) by each individual learner if it is to become an integrated part of the structure of knowledge held by the individual.

The view that knowledge must be constructed, or reconstructed, by each and every learner is currently attracting a great deal of interest. Lochhead (1985) outlined the view as follows:

> What I see as critical to the new cognitive science is the recognition that knowledge is not an entity which can be simply transferred from those who have to those who don't . . . Knowledge is something which each individual learner must construct for and by himself. This view of knowledge as an individual construction . . . is usually referred to as constructivism.

Thus knowledge is not a transferable commodity and communication is not a vehicle for effecting this transfer. The teacher's role is to help the learner in the conceptual organization and reorganization of experience, but it is the learner who must do the conceptualizing. In fact, not only does constructivism help us to understand the process of learning, it has implications for motivation too, according to von Glasersfeld (1987):

> if students are to taste something of the mathematician's satisfaction in doing mathematics, they cannot be expected to find it in whatever rewards they might be given for their performance but only through becoming aware of the neatness of fit they have achieved in their own conceptual construction.

It should be clear that the origins of constructivism would be difficult to trace, since the so-called Socratic method would seem to incorporate basically constructivist views about how learning takes place. Dienes (1960), in fact, in 'artificial motivation does not foster any love of the subject' and in speculating that exceptions to the otherwise gloomy picture regarding the learning of mathematics 'seem to occur when self-motivating learning-situations are created, where the information reaches the child in such a way that he can formulate it in his own language', came quite close to writing the same as von Glasersfeld. Indeed, in other statements such as 'mathematical learning being pre-eminently one of construction of predicates followed only afterwards by a critical examination of what has been constructed' and 'mathematical thinking will need the sort of investigation which catches the constructive process while it is going on', Dienes reveals his own version of constructivism. It is Piaget, however, who must be regarded as the most vital contributor in the development of contemporary constructivist views; indeed, reference has already been made to constructivism within Chapter 5. Kamii, in the titles of two books (1985, 1989) has described her work with children 'reinventing arithmetic' as 'implications of Piaget's theory'. Baroody (1987), in his description of current cognitive views of learning, even incorporates the word assimilation, thus forming a direct link with Piaget:

> Understanding is actively constructed from within by relating information to what is already known or by noticing a relationship between previously known but isolated pieces of information . . . Connecting new information to existing information . . . is called assimilation . . . New understanding can also occur by means of integration; connecting previously isolated bits of information.

The suggestion that learners must connect new information to already established knowledge structures and form new connections within and between structures is also reminiscent of ideas formulated by Ausubel and Novak. Constructivism is then, perhaps, a simple but profound expression of contemporary cognitive views of learning, having evolved naturally from earlier attempts to explain learning. Whether constructivism qualifies as a theory of learning is debatable. The most radical form of constructivism would, in fact, claim that 'we can never have access to a world of reality, only to what we ourselves construct from experience; all knowledge is necessarily constructed' (Goldin, 1989). Thus it could be claimed that it is impossible to ensure that any two learners have acquired the same knowledge, because each learner has constructed a unique model of reality. Such claims would seem to be within the domain of epistemology rather than learning theory.

The evolution of constructivism does not imply rejection of earlier attempts to facilitate better learning within a cognitive learning environment. It would seem to be a misunderstanding of constructivism to suggest that there is little the teacher can do to facilitate learning simply because the construction must be carried out by the learner. A discovery learning environment set up by the teacher might often be the best kind of environment. The provision of apparatus or manipulatives might be extremely important in providing an environment which enables children, particularly younger children, the better to reinvent. It is true that teachers have often been disappointed when the link between manipulatives and the arithmetic has not automatically been made by the child. This does not mean it is wrong, or even a waste of time, to provide the apparatus. Rather it suggests that perhaps even more is needed in terms of an environment in which to form the requisite link. Perhaps more discussion would help, between teacher and child and between child and child. Perhaps even that might not hasten the learning. Perhaps the child is not ready! But at least we can try to provide the best environment possible. Nor need we reject transmission out of hand. Although the emphasis in constructivism appears to be on spontaneous or 'bottom-up' learning, there is no need to reject the 'top-down' approach of interactions with knowledgeable adults or peers, particularly for older learners.

It is one thing to suggest that transmission is often ineffective but it would be unhelpful to suggest we must never use it. Ausubel has been insistent on this point. Being told something by a teacher or parent or child might be just what a learner requires at a particular moment in order to help construct meaning. Many children are capable of making new ideas their own quite quickly when these ideas are transmitted to them. If that were not the case, the whole of knowledge would need to be reconstructed by each individual, which is not only unnecessary, it is also a ridiculous proposition. In addition, much construction takes place subsequent to instruction, through reflection and even perhaps subconscious activity. Even repetitive chanting and drill exercises might promote construction, though construction would be unlikely if these were the only learning experiences provided for the child. What we must take care over is that we do not assume that only one method will lead to relational understanding through construction. The teacher needs to provide appropriate 'scaffolding' which allows the child to progress, and it requires great skill to provide the best scaffolding for each pupil. A consistent policy of complete non-intervention by the teacher is certainly not likely to be the best way to promote the reconstruction of knowledge. However, a policy of non-intervention with a certain child at a particular point in time or with a particular group might be appropriate, especially when the responsibility for learning has been fully accepted by the child or group, as might occur in the ideal open-learning or supported self-study scheme. Whether learning is basically active or passive is not the critical issue. What matters is whether the approach used has enabled the construction of meaning. As a generalization, active methods seem to be preferable for many children and for much of the time, but we certainly need much more knowledge about what methods best promote construction. Such knowledge, of course, may well be both topic-specific and child-specific! It must also be acknowledged that any interpretation of constructivism is itself a construction of the interpreter.

The work of Kamii is an example of an attempt to provide the kind of environment which best facilitates construction. One idea (Kamii, 1985) was to develop 'a first grade arithmetic program . . . by throwing out all traditional instruction and using, instead, only situations in daily living and games'. Another important aspect was social interaction, or rather 'the mental activity that takes place in the context of social exchanges'. Thus it is necessary for children to have to defend their point of view in front of their peers. 'If a child thinks that 8 + 5 = 12, he should be encouraged to defend his idea until he decides that another solution is better'. In a classroom in which such methods are adopted it would be common to witness children debating with each other, hearing contrary views to their own being propounded, having to justify their answer to a problem, and over a period of time perhaps changing their minds. It seems to be vital within the scheme that children decide for themselves when someone else's idea is better than their own in the belief that 'the nature of logico-mathematical knowledge is such that every teacher can be certain that children will arrive at correct answers, if they argue long enough among themselves'. It could be suggested, of course, that it is not 'correct' answers which will be accepted in this way, it will be those answers which are socially acceptable. In the realms of, for example, mechanics the sceptic might wonder whether certain 'correct' ideas would ever appear in such a debate between learners. The sceptic might also suggest that particularly sensitive or introvert children might find such teaching approaches unacceptable. Nevertheless, this deliberate application of what might be called a constructivist approach to teaching needs to be taken seriously, perhaps particularly with young children.

Another interesting set of experiments which appear to have similarities to the work of Kamii has been reported by Bell *et al.* (1989). The experiments were based on identifying and then eliminating particular misconceptions using a conflict-discussion procedure. The method demands good diagnostic test questions which it is known will reveal difficulties and misunderstandings, and much knowledge of such questions already exists in, for example, Hart (1981). On the basis of such a question a lesson in which pupils might record their own response, then discuss responses in small groups, and finally put the group conclusions to the whole class, on the assumption that the group work helps to ensure that pupils' wrong ideas are actually brought out and expressed, and that they can be subjected to challenge and criticism 'in an unthreatening situation'. Clearly, the unthreatening nature of the situations which is intended might be difficult to achieve for all individual pupils, and indeed in certain cultural environments. The stated aims of the research were 'to develop a way of teaching which contributes clearly to long-term learning and which promotes transfer'. The constructivist underpinning of this technique is captured in the statement 'the aim is that pupils should reach well-founded convictions based on their own perceptions, not take over superficially understood ideas from the teacher'. Early reports of the success of this technique are very encouraging. The most significant common feature between the methods described by both Kamii and Bell seems likely to be the emphasis on social interaction through discussion, debate and even argument. Clearly, much more research is needed.

A further experiment of great interest has been the Calculator Aware Number (CAN) Project, based at Homerton College, Cambridge under the direction of Hilary Shuard. This project was really set up as a part of the PrIME Project, which operated with the wider brief of providing a much more exploratory and investigatory approach to mathematics in the primary school. The basic intention of the CAN Project was to teach mathematics with a calculator at hand as much as possible, and not to artificially deprive children of what is a very powerful, useful and ubiquitous aid to calculation. In the English context this was a radical move and, quite naturally, advocates of proficiency in pencil and paper mathematics (many of whom are not teachers) have viewed what they would regard as a revolutionary development with suspicion. It is curious that there should be this lingering firm belief in the value of pencil and paper methods when so much of the Eastern world managed without such methods for so long, and when many developing countries do not have the resources to provide pencil and paper for their pupils and therefore have to manage without. The calculator can, of course, be used in an exploratory and investigatory way, and pupils are able to use the calculator to help in constructing their own understanding of arithmetic. If calculators are not built into the mathematics curriculum in a growing number of countries the chances are that the pupils will still use them at home, and will learn ahead of the teacher, and will also come to regard school mathematics as antiquated. The progress made by children taught within this scheme has been described by CAN (1990):

> Children seemed to be more confident, and more willing to explore mathematics for themselves than their predecessors. CAN teachers had developed their teaching in an investigatory and problem-solving direction. The children had not been taught the traditional vertical pencil-and-paper methods of carrying out addition, subtraction, multiplication and division, but had been encouraged to think out mathematical ideas for themselves. Consequently, unlike most primary children, the CAN children have not learned that there is a standard way of doing every mathematical process, and the teacher knows it and shows you how to do it. On the contrary, they have learned that in mathematics, you think things out for yourself.

INFORMATION-PROCESSING

It seems that any study which attempts to investigate and understand how information is processed in the mind can claim to be a part of that approach to the study of learning which has become known as 'information-processing'. It should therefore be clear that information-processing, as a theory of learning, is virtually impossible to describe in any simple way, because of the variety of studies which have been carried out under its umbrella. The concept of the mind as a processor of information has certainly led to research from many different perspectives. Cobb (1987) has claimed that 'historically, information-processing psychology developed as an alternative to the stimulus-response concept of behaviorism'; in particular, interest lay principally in what went on in the mind between stimulus and response. That may be true, but many more recent studies have little in common with behaviourism. In fact Cobb, in his paper, was attempting to look at information-processing from a constructivist perspective. Cobb quotes Sternberg in:

> That Piagetian theory is compatible with information-processing theory is shown by the fact that Rumelhart and Norman have proposed two modes of knowledge acquisition in information-processing language that correspond almost exactly to assimilation and accommodation.

Nevertheless, information-processing is often thought of as the other contemporary theory of learning apart from constructivism.

One impact of the electronic computer on education which is perhaps not widely appreciated is that contemporary theorists of human learning have frequently looked to the computer as a model of the human mind. Memory is seen to be the key to learning, for the objective is storage within, and ready recall from, long-term memory. The primitive diagram in Figure 9.3, which illustrates a simple interpretation of long-term memory, reveals the kind of connection, which many researchers have seen, with the model of how a computer works, with input, control, processing unit, store and output. The analogy with the computer has been taken further, in suggesting that the human mind has a built-in ready-for-action ROM (read-only memory) from the moment of birth.

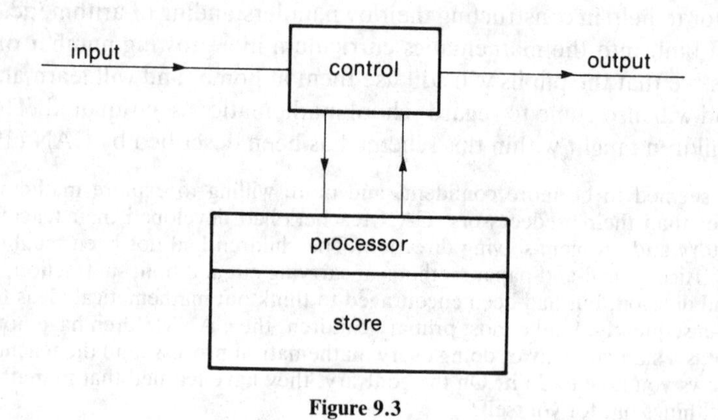

Figure 9.3

The process of committing knowledge to long-term memory, as suggested by Lindsay and Norman (1977), is illustrated in Figure 9.4. Consistency with Ausubel is

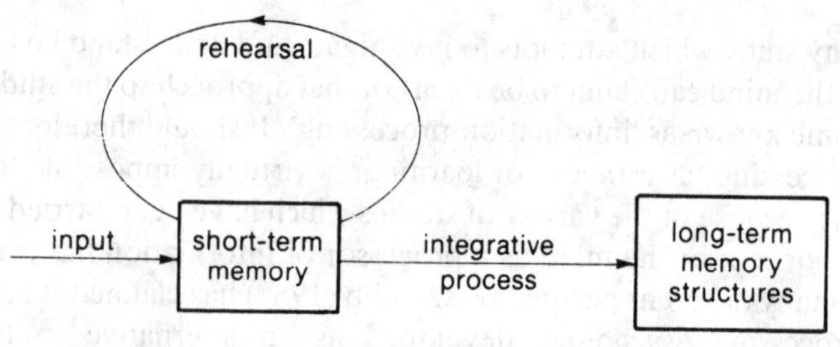

Figure 9.4

reflected in the references to 'integrative process' and 'memory structures'. A much more complex derivation from the elementary models shown here is to be found in Stewart and Atkin (1982). The work of Newell and Simon (1972) is often cited as being seminal in information-processing approaches to learning theory, and some of the many researchers who have trodden the same path, perhaps going even further, are referenced in Stewart (1985). A good study of cognitive development based on information-processing is that by Gross (1985). Readers might also consult the chapter by Davis (1983).

One outcome of cognitive science has been the continued study of semantic networks as representations of structures of knowledge stored in long-term memory. Semantic networks consist of nodes, representing concepts, linked by lines which express the relationship between the concepts, and these networks are otherwise known as concept maps. It was indicated earlier that the value of such maps has long been appreciated in the sciences, particularly biology, but that little use has been made of them in mathematics. Figure 9.5 is an example of a concept map which might

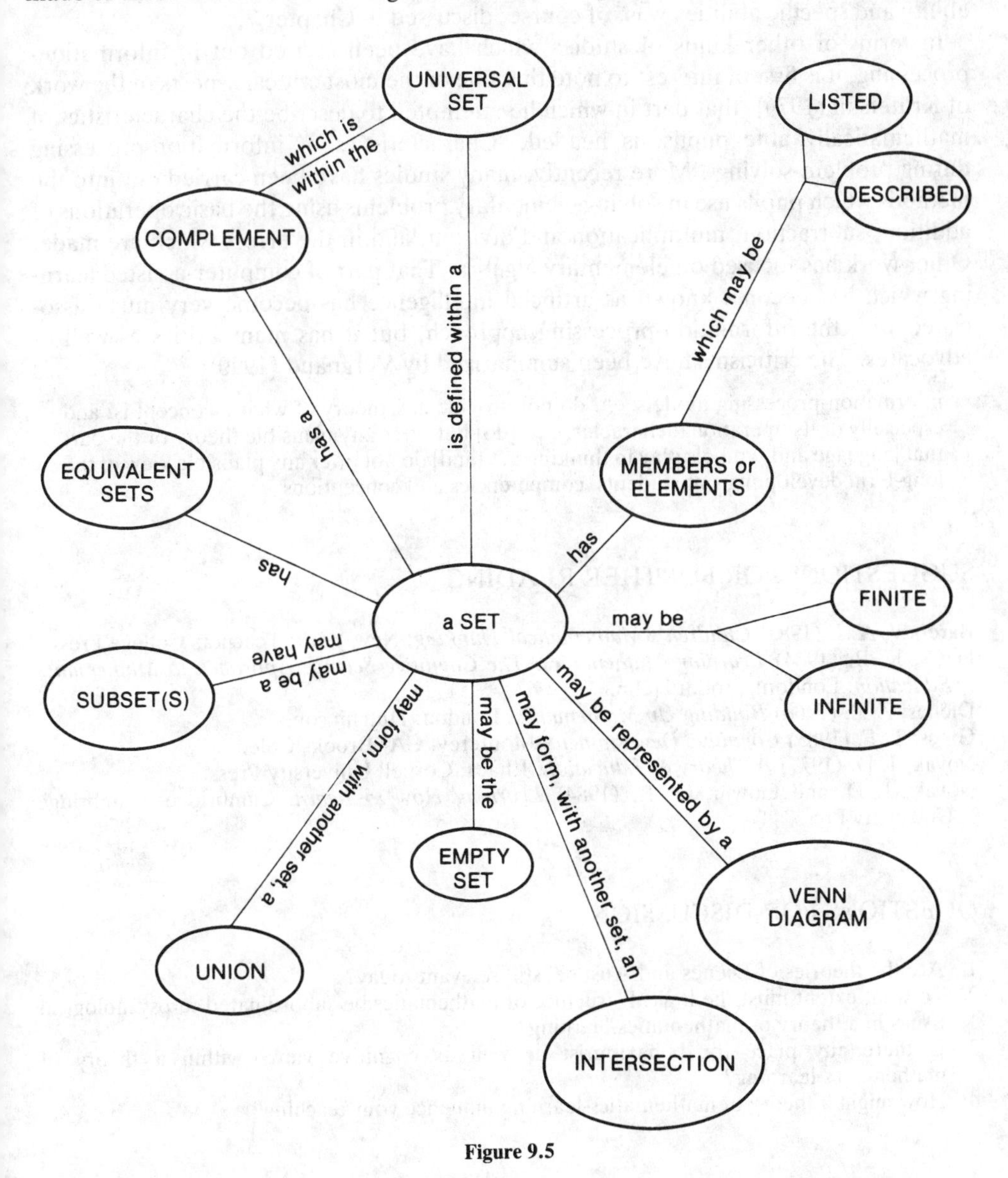

Figure 9.5

assist the teacher to teach and the learner to learn. There is, of course, a physiological equivalent to the psychological study of information processing, and interested readers might like to refer, for example, to Esler (1982).

Models of the mind which draw comparisons with how a computer works are very much a part of the contemporary scene in educational research, but Kilpatrick (1985) has suggested that we are also now 'witnessing a revival of faculty psychology', in that some theorists are now expressing the view that the human mind is like a collection of 'partially autonomous smaller minds' which have specialized functions and which operate in parallel. The issue of the relationship between overall ability, general ability and specific abilities was, of course, discussed in Chapter 7.

In terms of other kinds of studies which have been carried out in information-processing, it is first of interest to note that one of the most critical aspects of the work of Krutetskii (1976), that part in which he attempted to describe the characteristics of mathematically able pupils, is headed, 'Characteristics of information-processing during problem-solving'. More recently, many studies have been carried out into the methods which pupils use in solving elementary problems using the basic operations of addition, subtraction, multiplication and division, and in the errors which are made. Other work has focused on elementary algebra. That part of computer-assisted learning which has become known as artificial intelligence has become very much associated with the information-processing approach, but it has many critics as well as advocates. The criticisms have been summarized by Vergnaud (1990):

> information-processing models . . . do not provide any theory of what a concept is, and especially of its operational character, . . . [do] not offer any plausible theory of the part that language and symbols play in thinking . . . [and] do not offer any plausible view of the long-term development of students' competencies and conceptions.

SUGGESTIONS FOR FURTHER READING

Baroody, A. J. (1987) *Children's Mathematical Thinking*. New York: Teachers College Press.
Davis, R. B. (1984) *Learning Mathematics: The Cognitive Science Approach to Mathematics Education*. London: Croom Helm.
Dienes, Z. P. (1960) *Building Up Mathematics*. London: Hutchinson.
Gross, T. F. (1985) *Cognitive Development*. Monterey, CA: Brooks/Cole.
Novak, J. D. (1977) *A Theory of Education*. Ithaca: Cornell University Press.
Novak, J. D. and Gowin, D. B. (1984) *Learning How to Learn*. Cambridge: Cambridge University Press.

QUESTIONS FOR DISCUSSION

1. Are the theories of Dienes and Ausubel still relevant today?
2. To what extent must the logical structure of mathematics be subordinated to psychological issues in a theory of mathematics-learning?
3. Is there any place for behaviourist as well as cognitive views within a theory of mathematics-learning?
4. How might a theory of mathematics-learning influence your teaching?

Chapter 10

How Should Mathematics Be Taught?

The objective of teaching is to promote learning. Teaching often takes place, however, without learning being the outcome, and it is appropriate to consider whether teaching might be improved and learning optimized as a result of greater application of what is known about the process of learning. It is not widely acknowledged outside the profession of educators that learning is not a simple matter. If it were simple, we would all have adopted the elementary rules of teaching and our pupils would all be achieving great success. There is an element of uncertainty about how we learn, and considerable debate about the value of some theories, so rules for teaching are not easy to confirm. Nevertheless, there are aspects of what is known about learning which should be taken into account in teaching, because there is now fairly wide agreement. Some of these aspects have received mention in recent reports, such as Cockcroft (1982). This book is not, strictly, another of the many on teaching mathematics, but it is now appropriate to look briefly at how mathematics should be taught, in the light of what is known about learning, and to relate this to advice which is currently being given to teachers.

Clearly, learning units must be sequenced carefully. Even if we believe in the most extreme form of open learning possible, in which no formal teaching ever takes place and children are expected to discover knowledge from their learning environment, it is unlikely that sequencing can be completely ignored. We, the teachers, have to provide the environment and that environment will need careful planning. At the other extreme, the views of Skinner (1954) may not be acceptable for a number of reasons including the conviction that sequencing based only on stimulus-response connections is not appropriate for all forms and stages of learning and, what is more, it bores the pupils. Somewhere between the two extremes there is likely to be a widely acceptable attitude to sequencing, in which a logical progression of the subject of mathematics is laid before the pupils, or made available to the pupils, in a way that makes it possible for them to construct meaning. It might seem that there is nothing new here, yet emphasis on construction must imply that the sequencing takes account of the need for active involvement with materials and investigatory situations to a greater extent than has often been the norm. Also, emphasis on the development of knowledge structures implies careful thought about the relationships between concepts and between areas of knowledge.

A suitable, limited topic area to consider in brief outline from the point of view of sequencing, and learning generally, is graphical and pictorial representation of data. The essential graphical forms involved are:

- bar chart (and pictogram)
- line (or stick) graph
- jagged line graph
- histogram
- pie chart.

A concept map or similar analysis would reveal, however, that there are many underlying and associated ideas, for example:

- tallying, tables
- scale, origin
- axis, axes
- variable
- independence, dependence
- coordinates
- discrete, continuous
- frequency, distribution.

These ideas have all been discussed, from the point of view of the desired ultimate objectives and the difficulties which might be encountered on the way, in Carter *et al.* (1981). Although the ideas might seem, at first sight, to be easy, there are a number of problems of learning which need to be acknowledged. One problem arises with pie charts because of their association with ideas from outside the above lists, namely angle and proportion. Other problems arise with histograms, both with their construction from data collected in the field which needs to be grouped and also with the slightly hazy overlap area with bar charts in, for example, probability experiments.

The early stages of work on graphical representation should, however, be fairly straightforward. These early stages were reviewed in Nuffield (1967b), and many of the pitfalls were indicated, though the possibility of cognitive dissonance later, if care was not taken over two particular aspects, was not, apparently, realized, and these aspects will be mentioned later. Clearly, simple concrete beginnings are essential to the construction of knowledge so real objects must be used at first. These objects need to be in one-to-one correspondence with what is being represented, so scale must not be introduced yet. Only two rows or columns should be used as, for example, in comparing children who have brothers and/or sisters in the school with children who do not. Subsequently, a return to these elementary ideas from time to time throughout a child's education allows extension and refinement in the manner of the spiral curriculum of Bruner (1960b), so that more rows or columns than two can be compared and scaling can be introduced. The spiral approach also allows the initial use of actual objects to be replaced by pictures of objects and finally by a coloured square or a symbol, in other words the enactive → iconic → symbolic sequence of Bruner (1966) should be used. It is interesting that this is one feature of planning learning which was directly referred to in Cockcroft (1982):

> and so to progress within each topic from the handling of actual objects to a stage in which pictures or diagrams can be used to represent these objects and then to a final stage at which symbols are used which can be manipulated in abstract ways.

These early graphs are often referred to as 'block graphs', because the lines usually become lines of squares eventually, with or without pictures in them. Block graphs themselves are, in fact, only a stage on the way to the usual forms of graphical representation so that, by progressive differentiation, discrete data for which the independent variable is non-numerical becomes represented by a bar chart, discrete data for which the independent variable is numerical becomes represented by a line (stick) graph, and continuous data becomes represented by a jagged line graph, or by a histogram if the data is grouped (see Figures 10.1 to 10.4).

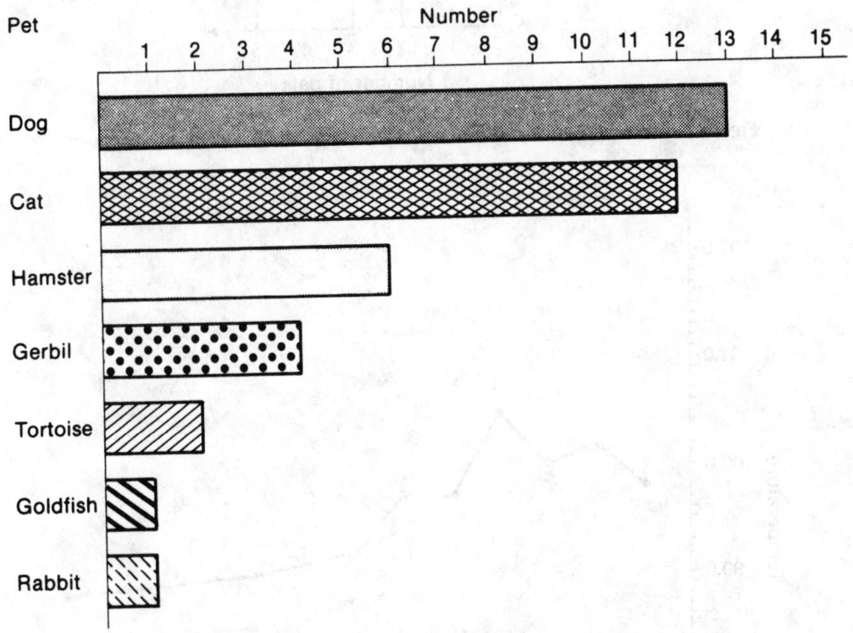

Figure 10.1 *Bar chart showing numbers of each pet owned by Class 3.*

The fact that the block graph is the foundation for most forms of representation assigns it enormous importance. It is vital that the opportunity to set children on the right road is grasped, even as early as the infant stage. In some texts the lines of blocks are placed directly adjacent, but this is open to criticism because it is not appropriate for discrete data. In both the bar chart and the line (stick) graph the lines should be separated by some uniform gap, so if unnecessary cognitive dissonance is to be avoided it could be argued that these gaps should be there right from the very beginnings of block graphs. There could, of course, be exceptions if the independent variable is continuous, for example in graphing the months of the year (independent) against the number of children with birthdays in each month (dependent), though children might then wonder why some graphs were allowed to have adjacent lines of blocks and others not. It is also important, at this early stage, to organize the blocks into lines in both perpendicular directions, both as columns and as rows. Line (stick)

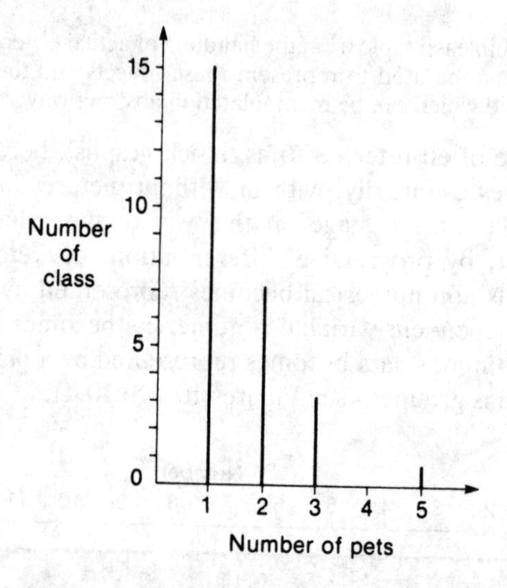

Figure 10.2 *Graph showing how many members of Class 3 own how many pets.*

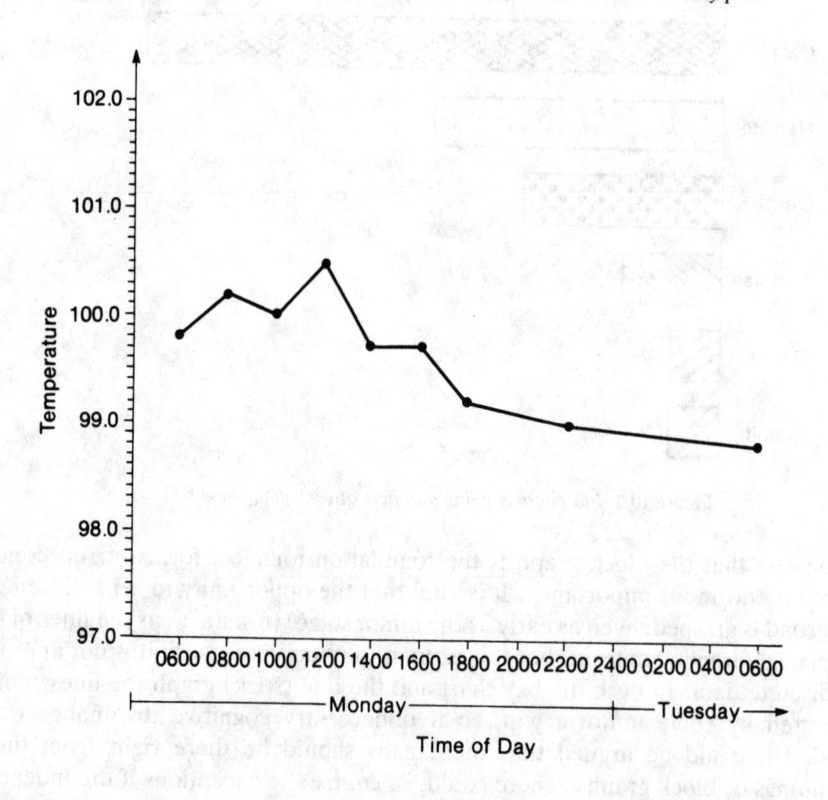

Figure 10.3 *Graph showing Gerard's temperature in hospital.*

graphs are conventionally of column form whereas bar charts and pictograms are more frequently arranged in row form. There is also the problem of how to describe the two perpendicular directions for the axes. Nuffield (1967b), along with most other

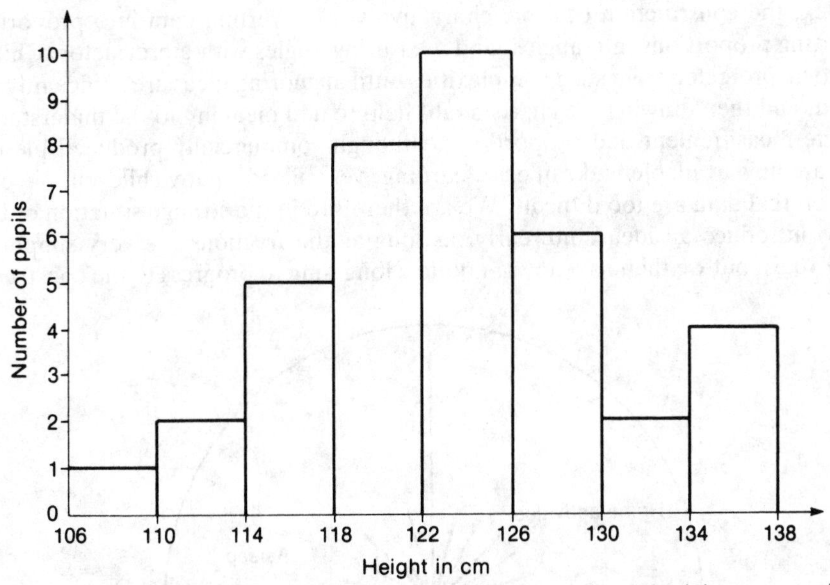

Figure 10.4 *Graph showing distribution of heights of pupils in Class 4.* (Note that it is not appropriate here to discuss details of histogram construction. Decisions concerning how many groups of heights and what are the boundaries between groups need to be taken with care. Further details will be found in many books, for example, Carter *et al.* (1981) and Hodge and Seed (1972).)

sources, referred to them as horizontal and vertical. How we have got into a situation in which lines which are often not vertical are described as vertical is a mystery, and there is certainly a potential problem of cognitive dissonance here which needs careful handling. The problem is eventually solved by '*x*' and '*y*'; prior to that some authors use 'across' and 'up' (!) the page.

At some stage, coordinates will be introduced. There are a number of useful activities which can help to provide both motivation and a concrete beginning, including arranging the children in parallel rows and columns and assigning coordinates to each child, hunting for imaginary buried treasure, using one of the many computer programs, and plotting shapes by joining dots. Once coordinates have been introduced the idea feeds into the development of graphical ideas in both line graphs of discrete data and jagged line graphs of continuous data. It is then very important that discrete data which should be represented by a line (stick) graph is never joined up to form a continuous jagged line graph, which would incorrectly suggest that there is meaning to intermediate points on the jagged line. Many jagged line graphs will have 'time' as the independent variable, but there are other possibilities such as conversion graphs (which are clearly not *jagged* though they are of the same type).

Pie charts present a particular problem of readiness. Simple pie charts are very simple, but complicated pie charts are very difficult indeed! However, these particular graphical forms are very appealing and are in widespread use in the real world, so children have to meet them. The issue of sequencing which is raised is that sequencing frequently has to take account of what the child already knows not only within the particular topic area, representation of data in this case, but also in other topic areas. Unfortunately, real-life data rarely produces simple proportions and, at its most complex, the construction of a pie chart involves converting data into proportions, converting proportions into angles, and measuring angles with a protractor. Children have to be protected from such complexities until anchoring ideas are sufficiently well-formed, and then drawing pie charts might help to add meaning to the understanding of angle measurement and proportion. Although commercially produced pie chart scales are now available and can ease learning, yet still, for many children, pie charts based on real data are too difficult. We are therefore in the strange situation of being able to introduce an idea quite early, as long as the fractions are very simple (see Figure 10.5), but we then have to wait quite a long time to progress to the construction

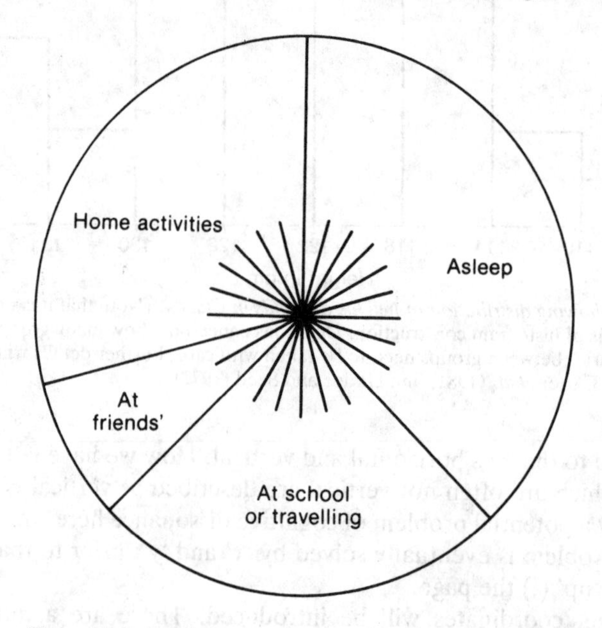

Figure 10.5 *Graph showing how Debbie spends her weekdays.*

of pie charts from collected data. Sequencing must take account of inherent difficulties in the nature of the subject of mathematics, like proportionality.

This brief consideration of certain aspects of the representation of data in graphical form has raised some of the issues but cannot be a complete study of how to teach graphs; an entire book would be required for that. Some of these issues, however, are worthy of further elaboration through illustrations of other mathematical topics. How, for example, should we help the child to learn that the area of a circle is πr^2, where r is the radius? Several elegant methods which are frequently suggested, even sometimes in primary school books, involve the idea of limit. For example, if a circle is divided into sectors and the sectors are then rearranged in a line (see Figure 10.6), and the process is taken to a limit so that we have more and more sectors with smaller and smaller angles at the centre, we may eventually appreciate that the required area is the same as for a rectangle of length πr and breadth r. But is this method meaningful to the child? Such is the way we ignore the important idea of limit in the mathematics curriculum that there is no way we can assume understanding of the limiting process when this method is used to introduce the idea of area of a circle. That is not to say

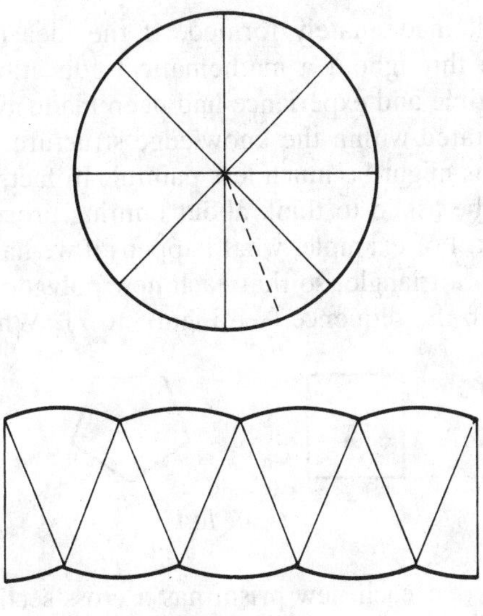

Figure 10.6

that children will gain nothing from such a method, for they might gain a great deal. However, it would seem much more sensible to use a more concrete approach like counting squares, which will yield a very good answer and can *subsequently* be used to talk about limits in connection with smaller and smaller squares. There is a strong sense in which the issue of sequencing suggests that the idea of limit is more advanced than the idea of area, so limits should not be used to teach area but area may be used to teach limits. Of course, this is not clear-cut, and a different curriculum with more emphasis on the idea of limit from the very earliest years might make the approach to πr^2 illustrated above much more acceptable.

The issue of the concept of limit within mathematics leads naturally to the next point. Even if optimum sequencing could be achieved, the next ideas suggested by careful structuring of the subject matter might not be learned because not enough time had elapsed to allow sufficient reflection on earlier ideas. The Schools Council Bulletin (1965) stated that children learn mathematical concepts more slowly than we had previously realized and that they learn by their own activities. The work of Piaget, of course, did much to bring this view to light. Dienes, as a result, laid great emphasis on activity and variety of experience over a period of time in order to promote learning. Cockcroft (1982) declared: 'Failure to understand a mathematical topic can result from . . . teaching which . . . moves ahead too rapidly so that understanding does not have time to develop', thus, 'Young children should not be expected to move too quickly to written recording in mathematics.' In terms of learning about limiting processes, how can we call upon ideas of limits if they have not been developed throughout the child's education in such a way as to allow ideas to become fully integrated within the knowledge structure?

When calculus is introduced to pupils it is necessary to refer to the idea of limit, yet this might be the very first encounter with limits in the experience of the students. In the case of calculus and in the case of using limits to introduce the area of a circle there is the danger that we might be trying to build on knowledge and understanding of a concept which is itself inadequately formed. If the idea of limit were developed whenever appropriate throughout a mathematical education there would be more chance that sufficient time and experience had been made available for real meaning to have become integrated within the knowledge structure. Then, for example, the introduction of calculus might be much less painful. In fact, there are many ways in which children might be asked to think about limiting processes in connection with more elementary work. For example, what happens if we have a sequence of regular polygons, starting with a triangle, so that each new polygon has one more side than the previous polygon in the sequence (see Figure 10.7)? What happens if we have a

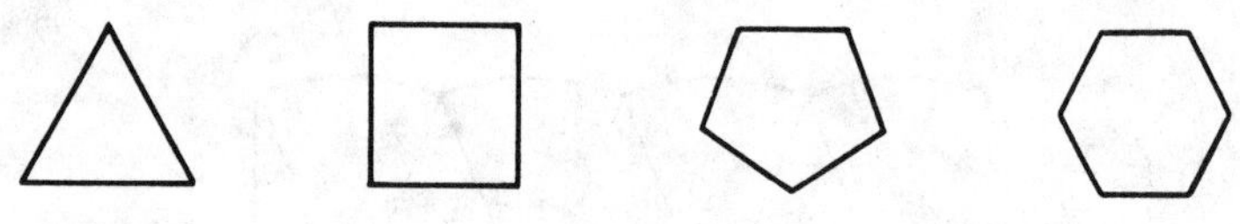

Figure 10.7

sequence of prisms so that each new prism has a cross-section with one more edge than the previous one? What happens with the equivalent sequence of pyramids? In each case the ideas of a never-ending process and of limit may be discussed. Simple number sequences and series are also easy to introduce, such as patterns of fraction sequences:

$$\tfrac{1}{2}, \tfrac{2}{3}, \tfrac{3}{4}, \tfrac{4}{5}, \tfrac{5}{6}, \ldots$$

$$\tfrac{1}{2}, \tfrac{1}{4}, \tfrac{1}{8}, \tfrac{1}{16}, \tfrac{1}{32}, \ldots$$

$$\tfrac{2}{2}, \tfrac{3}{4}, \tfrac{4}{6}, \tfrac{5}{8}, \tfrac{6}{10}, \ldots$$

Corresponding sums of series, for example

$$\tfrac{1}{2} + \tfrac{1}{4} + \tfrac{1}{8} + \tfrac{1}{16} + \tfrac{1}{32} + \ldots,$$

may easily be studied using strips of paper (see Midlands Mathematical Experiment, 1970). Plotting graphs of lines and curves involves a limiting process, for the more points which are plotted the closer we are to the best representation of the complete relationship. Carrying out probability experiments and collecting the results from around the class may be used to introduce the idea of limit. In the above, and in many other ways, limits may become a real part of the mathematics curriculum, so that appropriate concepts are available when some other idea in the sequence of mathematical experiences requires them. Ultimately, for those who continue with mathematics beyond school, limits might become very abstract in terms of symbolism and manipulation, but such a level of abstraction is not necessary when the idea of limit is being introduced.

Some mathematics, however, is quite abstract for young pupils. Algebra is the obvious difficult major branch of mathematics, abstracted from more concrete situations like relationships and patterns with numbers. The concrete/abstract issue is certainly an important one in mathematics-learning and must be taken into account in decisions about how mathematics should be taught. Piagetian stage theory has been used as a reference in discussing the relative difficulties of concrete and abstract (formal) tasks, and it is important that the debate about the relevance and acceptability of Piagetian theory must not be allowed to obscure the issue of concrete versus abstract approaches to mathematics. Even the early report of the Mathematical Association (1923) allowed for this kind of issue in its recommendations for teaching geometry through Stage A (concrete activity), Stage B (deductive) and Stage C (systematizing). There are, of course, no absolutes in the concrete/abstract issue, there are degrees of abstractness, and yesterday's abstract idea may become tomorrow's concrete example. Nevertheless, the issue needs to be taken into account more than it has been, for Cockcroft (1982) stated: 'A premature start on formal written arithmetic is likely to delay progress rather than hasten it', and also, 'At the present time up to 80 per cent of pupils in secondary schools are following courses . . . whose syllabuses are comparable in . . . conceptual difficulty with those which twenty years ago were followed by only about 25 per cent of pupils'. Some of the increased abstractness, which led to conceptual difficulties, was introduced with curriculum development in the 1960s and 1970s. One example is the topic of relations and functions, and the difficulties in this branch of mathematics have been discussed by Orton (1971).

One advantage of a spiral curriculum, in which topics are returned to with a view to development and extension, is that each new visit to the topic allows a different approach with a steady shift along a concrete/abstract continuum. The early work on graphical representation of data illustrates this point, with frequent returns to the topic of block graphs in order to introduce greater complications of scaling and also to move through the enactive and iconic stages to different symbolic representations. The same kind of progression could be made to work with functions, but examination pressures appear to have persuaded many teachers to aim for greater abstraction more quickly than pupils can absorb it. In the last resort, decisions about the level of abstraction incorporated in the work of a lesson are inevitably made by the teacher, often on a fairly intuitive basis. General statements about learning and theory provide advice of a very broad nature and cannot be child/topic/lesson specific. Detailed studies of children's understanding of mathematical topics are continuing to add to our knowledge and, whether general theories provide guidance or not, it is helpful for teachers to refer also to such detailed studies as Hart (1981).

Another approach to attempting to improve the quality and quantity of learning must be to try to integrate and then to extract the consensus from the many empirical studies which have been carried out in relation to each aspect of the mathematics curriculum. One important area is clearly the four operations (rules) of arithmetic, beginning with addition and subtraction. Here, an enormous body of research evidence now exists. Romberg (1982) has, on the one hand, stated that 'the many studies on addition and subtraction represent an eclectic morass', but on the other hand claims an emerging 'firm research consensus in this area'. It seems that teachers of young children should take note. One early study by Brownell (1947) clearly demonstrated that 'decomposition' is superior to 'equal addition' as a subtraction strategy, when taught with rational explanation, yet even this is not yet being implemented by all teachers.

The many studies referred to by Romberg really consist of much more recent work, such as that by Carpenter and Moser (1982). An important conclusion of this research, and of that by MacNamara (1990), seems to be that strategies being used by teachers of young children may run counter to the informal knowledge which pupils bring to the learning situation, and may therefore cause greater disequilibrium than is either necessary or wise. Carpenter and Moser claim that school mathematics programmes assume that addition and subtraction are best introduced through combining and separating sets, but children prefer to use basic counting strategies. For addition, three basic counting strategies were observed, namely counting all with models (e.g. fingers), counting all without models and counting on (either from the first number or from the larger number). In general, most children seemed to move on fairly quickly from counting all with models to counting on, with this latter strategy apparently becoming important when totals begin to exceed ten. Eventually, children acquire factual knowledge (number bonds) and are able to use these both in addition and to derive other facts, usually based on doubles or on numbers whose sum is ten. Thus 6 + 8 may be calculated by using 2 more than 6 + 6 (a known fact) or by using 4 more than 6 + 4 (another known fact). Piaget did not seem to accept counting as evidence of true assimilation of knowledge, apparently dismissing counting techniques as rote processes. Yet there is now evidence available that children can construct informal procedures for addition solely on the basis of counting. The view put forward by Starkey and Gelman (1982) is in severe contrast to that of Piaget. They claim that number forms a natural cognitive domain, that it is a 'cultural universal', and that coming to know about number is much like coming to know about language. It was Chomsky who postulated that we were all born with an inbuilt Language Acquisition Device which enabled us to construct language (speech). The suggestion of Starkey and Gelman is that we were also all born with an inbuilt set of counting principles. If that is the case, counting becomes the natural way of beginning to learn addition, and hence the rest of elementary arithmetic.

More importantly, however, from the point of view of implementation in the classroom, it is necessary to give serious consideration to what else teachers might do in order to enhance learning. Clearly, counting is a natural approach for children to use, and it seems it is likely to be counter-productive with many children to attempt to replace this with other techniques such as combining and separating. Counting on (from larger, eventually) needs to be encouraged when the child is ready. For addition, the order of difficulty for children seems to be 'doubles', followed by 'close numbers', followed by other sums. It seems sensible, therefore, to approach the addition of 'close numbers' by using doubling. The ability of even young children to 'subitize' is not being used (MacNamara, 1990). Subitizing is the ability to recognize how many objects there are in a group without counting them, and it seems that most children commencing school have this capability for groups of five or fewer. At school, these same children would be likely to be taught techniques which did not acknowledge subitizing, and which would therefore run counter to their natural inclinations, that is, until the school had stamped out this more intuitive approach. If we place young children, just beginning school, in a situation in which their mathematics demands that they cease to use effective capabilities they already have, then mathematics gets off to a bad start from which it will perhaps never recover. Ascertain what children know and teach accordingly, is the long-standing recommendation of Ausubel which we seem to find difficult to implement. But in addition, teachers will need a very clear indication of what is the firm consensus which Romberg has claimed is emerging.

The very great differences in mathematical attainment between children of the same age are now well documented and were summed up in the reference to a seven-year-difference centred on age 11 in respect of a particular place-value task (Cockcroft, 1982, from Brown, 1981b). Such wide differences have emerged from a wealth of studies conducted over many years and should not come as a surprise. There might be distinctive features of mathematical ability, but it is perhaps better to regard mathematical ability as one facet of overall ability which, however it has been measured over the past 60 years, has always produced a wide spread at any one particular age, giving rise to such concepts as mental age and intelligence quotient. The spread is certainly very noticeable in the context of mathematics, but it cannot be unique to mathematics. It is also clear that the gap between the highest and lowest attainers grows wider throughout the years of schooling. The implications for teaching are enormous. The HMI handbook of 1979 posed the question, 'What range of understanding and which skills should be acquired by pupils by the time they are 11?', without at the same time drawing attention to the extent of the spread of attainment at that age and the implications. Cockcroft (1982) referred to the problem very explicitly in saying: 'it is not possible to make any overall statement about the mathematical knowledge and understanding which children in general should be expected to possess at the end of the primary years.' This is a very great complication for teachers who might like to have clearly in mind what is the extent of their mathematical curriculum. One implication of the great difference in attainment at any age is that it may not only not be appropriate to provide the same work for all pupils of that age, but also it might be necessary to provide work at different levels of abstraction. Individualized or small group work might be the only way to ensure that the weak are allowed sufficient time and the very able are allowed to proceed more quickly.

Cockcroft (1982) also drew attention to the fact that the Committee received submissions which 'urged that more emphasis should be placed on "rote learning"'. With all the knowledge we now have about rote learning, summed up by the discussion in Ausubel (1968), this was amazing. The inadequacies of rote learning as the sole approach to learning, say, multiplication tables is now generally acknowledged, and the desirability of looking for meaning in all that is learned is agreed. Perhaps there is confusion in the minds of some people between justifiable rehearsal and pure rote. Certainly, for the memory to work efficiently, rehearsal and revision are important, but knowledge is retained better if it can be stored as part of a network of knowledge. Learning purely by rote cannot be defended because knowledge acquired in this way can only be stored in long-term memory separate from and unconnected with other knowledge, which makes both retention and recall difficult. Skemp (1976) raised the associated issue of instrumental and relational understanding. The issue of instrumental learning is rather more complex than the issue of learning by rote, for teachers do try, wherever possible, to teach algorithms for relational understanding. Pupils, however, often only retain the procedure and not the meaning. The conclusion we should draw for teaching is that we should look very carefully at those algorithms which we know are only understood instrumentally and decide whether they are really necessary and, if so, whether they should be developed as early as they often are. It would seem that the long division algorithm is very difficult to justify now, since pupils do not understand it and no longer need it. Long multiplication is also harder to justify now than it once was, but the distributive law is important in mathematics so some multiplication is necessary. Most of the fraction algorithms are now very difficult to justify at all, let alone in the primary school. An algorithm like the denary → binary conversion, explained in Chapter 3, cannot be justified at all, unless relational understanding can be achieved, which is most unlikely.

Within the logical development of mathematics there also appear to be important topics which are much more difficult than expected. Rees (1973) discovered that a wide variety of different groups of students found the same mathematical topics particularly difficult, and that the rank order of difficulty of a set of tasks was remarkably constant across these student groups. The topics naturally included ratio and proportion, but there were many more. Many teachers find ratio and proportion very difficult to teach, and those textbooks which base the work almost exclusively on finding the missing number from two equal ratios do not help. Some discussion of situations in which ratio is the underlying structure is essential if ideas are to be allowed time to develop. Some of the research tasks invented to test pupils' understanding are, in fact, appropriate vehicles for this purpose. We are now much more aware of the need for more genuine discussion in mathematics lessons to allow pupils to exchange and thereby to refine their ideas whilst the teacher plays a much more distant and subordinate role than in the traditional lesson of exposition with some questions and answers. Hart (1984) has also provided us with some insight into how teaching might be specifically geared towards the eradication of particular, common and known misconceptions.

Knowledge about how pupils learn mathematics and why they often fail to learn mathematics continues to accumulate. Fine detail is provided by every study, no matter how restricted and 'microscopic' it might be, in the sense of looking at only a very limited section of mathematics. It is easy to overlook the fact that our highest attainers possess enormously extensive and detailed knowledge of ideas which contribute to their understanding of mathematics by the time they leave school. The work of research into levels of understanding of the minutiae of knowledge possessed by children will go on. At the other extreme, more general theoretical views are being developed from previous general theories. These details and prevailing theories are certainly now reflected in official reports, like Cockcroft (1982), which included a more extensive discussion of problems and issues of learning mathematics than has been seen in any previous report. The summary of good practice in mathematics teaching, included in paragraph 243 of this report, has its origins in many disparate sources but certainly owes something to what is known about issues of learning. Previous reports have also included some statements about learning, and it is a sad fact that classroom practice often appears to suggest ignorance of the kinds of recommendations which have been made over many years. Plowden (1967) stated that:

> Instruction in many primary schools continues to bewilder children because it outruns their experience . . . even in infant schools, where innovation has gone furthest, time is sometimes wasted in teaching written 'sums' before children are able to understand what they are doing.

The Schools Council Bulletin (1965) declared that there was 'unchallengeable evidence that sound and lasting learning can be achieved only through active participation'. Hadow (1931) included the statement that: 'The curriculum is to be thought of in terms of activity and experience rather than of knowledge to be acquired and facts to be stored.' McIntosh (1977) referred to *HMI Reports* of 1895 which included:

'no instruction in the rules of arithmetic can be really valuable unless the process has been made visible to the scholars by numerous concrete examples'. Most of the above statements were made in relation to primary school children, but more recently it has come to be appreciated that similar statements can be made in terms of secondary school pupils and their mathematics. Amongst many statements about older pupils Cockcroft (1982) declared: 'It is too often assumed that the need for practical activity ceases at the secondary stage but this is not the case . . . nor is it the case that practical activity is needed only by pupils whose attainment is low.' Hart (1981) stated: 'different embodiments, more concrete referents must be brought in all the time.' Even that very select group of mathematics students in the 16–21 age range experience difficulties and require practical, concrete approaches, which allow discussion, as has been revealed in Orton (1980, 1983a, 1983b, 1985).

Views on how children and students learn mathematics have, in fact, reached the stage where certain beliefs have become consistent. For example, emphasis is now placed on the extended use of concrete experiences and a more gradual move to abstraction, on the support of relevant practical activities, on the use of apparatus or equipment, on the unsatisfactory and ineffective nature of learning by rote, on the importance of integrating knowledge in a meaningful way, on the value of discussion and on the need to cater for individual differences. These views are all aspects of learning which have been discussed throughout this book and which are currently included in advice to teachers in terms of classroom practice. As regards learning theory, an eclectic approach commends itself, such that within an environment which encourages the construction of knowledge and understanding carefully sequenced elements of learning experience can exist. In other words, cognitive views perhaps determine the prevailing climate within which behaviourist approaches to education need not be rejected out of hand. From the point of view of the child, variety is important anyway.

SUGGESTIONS FOR FURTHER READING

At this stage in a book of this nature it is tempting to include all those recent books which are worth reading and which have not so far been recommended. The list below is a compromise, but still covers a very wide field.

Bell, A. W., Costello, J. and Küchemann, D. (1983) *Research on Learning and Teaching*. Windsor: NFER–Nelson.

Cambridge Institute of Education (1985) *New Perspectives on the Mathematics Curriculum*. London: DES.

Choat, E. (1978) *Children's Acquisition of Mathematics*. Windsor: NFER.

Farrell, M. A. and Farmer, W. A. (1980) *Systematic Instruction in Mathematics for the Middle and High School Years*. Reading, MA: Addison-Wesley.

Ginsburg, H. (1977) *Children's Arithmetic: The Learning Process*. New York: Van Nostrand.

Ginsburg, H. P. (ed.) (1983) *The Development of Mathematical Thinking*. New York: Academic Press.

Hughes, M. (1986) *Children and Number*. Oxford: Basil Blackwell.

Larcombe, A. (1985) *Mathematical Learning Difficulties in the Secondary School*. Milton Keynes: Open University Press.

Lesh, R. and Landau, M. (1983) *Acquisition of Mathematics Concepts and Processes*. New York: Academic Press.

Resnick, L. B. and Ford, W. W. (1984) *The Psychology of Mathematics for Instruction*. Hillsdale: Lawrence Erlbaum.

QUESTIONS FOR DISCUSSION

1 In what ways have you changed your views about the most effective ways of promoting the learning of mathematics as a result of reading this book?
2 What are the most important issues concerned with learning mathematics with which all mathematics teachers should be acquainted?
3 How is it possible to enable all children to learn what is the real nature of mathematics?
4 What evidence is there that today's children are benefiting from our accumulated knowledge about learning mathematics?

Bibliography

Assessment of Performance Unit (1982a) *Mathematical Development: Primary Survey Report Number 3*. London: HMSO.

Assessment of Performance Unit (1982b) *Mathematical Development: Secondary Survey Report Number 3*. London: HMSO.

Austin, J. L. and Howson, A. G. (1979) Language and mathematical education. *Educational Studies in Mathematics* **10**, 161–97.

Ausubel, D. P. (1960) The use of advance organizers in the learning and retention of meaningful verbal material. *Journal of Educational Psychology* **51**, 267–72.

Ausubel, D. P. (1963) Some psychological and educational limitations of learning by discovery. *New York State Mathematics Teachers Journal* **XIII**, 90–108. (Also in *The Arithmetic Teacher* **11** (1964), 290–302.)

Ausubel, D. P. (1968) *Educational Psychology: A Cognitive View*. New York: Holt, Rinehart & Winston.

Barker, K. (1979) Place-value in children aged 7–9 years. Report of Teacher Fellowship, University of Leeds.

Barnes, D. (1976) *From Communication to Curriculum*. Harmondsworth: Penguin Books.

Barnes, D. (1985) The role of verbalizing in classroom learning. In A. Orton (ed.), *Proceedings of the 1985 Conference of the British Society for Research into Learning Mathematics*. University of Leeds Centre for Studies in Science and Mathematics Education.

Baroody, A. J. (1987) *Children's Mathematical Thinking*. New York: Teachers College Press.

Bartlett, F. (1958) *Thinking*. London: George Allen & Unwin.

Bell, A. W., Costello, J. and Küchemann, D. (1983) *Research on Learning and Teaching*. Windsor: NFER–Nelson.

Bell, A., Wigley, A. and Rooke, D. (1978–9) *Journey into Maths: Teacher's Guides*. Glasgow: Blackie.

Bell, A. *et al.* (1989) *Diagnostic Teaching*. Nottingham: Shell Centre for Mathematical Education.

Bell, P. (1970) *Basic Teaching for Slow Learners*. London: Muller.

Benbow, C. P. and Stanley, J. C. (1980) Sex differences in mathematical ability: fact or artifact? *Science* **210**, 1262–4.

Berry, J. W. (1985) Learning mathematics in a second language: some cross-cultural issues. *For the Learning of Mathematics* **5**(2), 18–23.

Bigge, M. L. (1976) *Learning Theories for Teachers* (4th edn). New York: Harper & Row.

Biggs, E. E. (1972) Investigational methods. In L. R. Chapman (ed.), *The Process of Learning Mathematics*. Oxford: Pergamon Press.

Biggs, J. B. (1962) *Anxiety, Motivation and Primary School Mathematics*. London: NFER.

Bishop, A. J. (1973) The use of structural apparatus and spatial ability—a possible relationship. *Research in Education* **9**, 43–9.
Bishop, A. J. (1980) Spatial abilities and mathematics education—a review. *Educational Studies in Mathematics* **11**, 257–69.
Bishop, A. (1988) Mathematics education in its cultural context. *Educational Studies in Mathematics* **19**, 179–91.
Bloom *et al.* (1956) *Taxonomy of Educational Objectives: Cognitive Domain*. London: Longman.
Booth, L. R. (1984) *Algebra: Children's Strategies and Errors*. Windsor: NFER–Nelson.
Branford, B. (1921) *A Study of Mathematical Education*. Oxford: Clarendon Press.
Brighouse, A., Godber, D. and Patilla, P. (1982) *Peak Mathematics 5*. Walton-on-Thames: Nelson.
Brissenden, T. (1988) *Talking About Mathematics*. Oxford: Blackwell.
Brown, G. and Desforges, C. (1977) Piagetian psychology and education: time for revision. *British Journal of Educational Psychology* **47**, 7–17.
Brown, M. (1978) Cognitive development and the learning of mathematics. In A. Floyd (ed.), *Cognitive Development in the School Years*. London: Croom Helm.
Brown, M. (1981a) Number operations. In K. M. Hart (ed.), *Children's Understanding of Mathematics: 11–16*. London: John Murray.
Brown, M. (1981b) Place-value and decimals. In K. M. Hart (ed.), *Children's Understanding of Mathematics: 11–16*. London: John Murray.
Brownell, W. (1947) An experiment on 'borrowing' in third-grade arithmetic. *Journal of Educational Research* **41**(3), 161–71.
Bruner, J. S. (1960a) On learning mathematics. *The Mathematics Teacher* **53**, 610–19.
Bruner, J. S. (1960b) *The Process of Education*. Cambridge, MA: Harvard University Press.
Bruner, J. S. (1966) *Toward a Theory of Instruction*. Cambridge, MA: Harvard University Press.
Bruner, J. S. (1973) *Beyond the Information Given*. London: Allen & Unwin.
Bruner, J. S., Goodnow, J. J. and Austin, G. A. (1956) *A Study of Thinking*. New York: Wiley.
Bruner, J. S. and Kenney, H. J. (1965) Representation and mathematics learning. In L. N. Morrisett and J. Vinsonhaler (eds), *Mathematical Learning*. Monograph of the Society for Research in Child Development **30**(1).
Bruner, J. S., Olver, R. R. and Greenfield, P. M. (1966) *Studies in Cognitive Growth*. New York: Wiley.
Bryant, P. (1974) *Perception and Understanding in Young Children*. London: Methuen.
Burton, L. (1984) *Thinking Things Through*. Oxford: Blackwell.
Buxton, L. (1981) *Do You Panic About Maths?* London: Heinemann.
Byers, V. and Erlwanger, S. (1985) Memory in mathematical understanding. *Educational Studies in Mathematics* **16**, 259–81.
Calculator Aware Number Project (1990). *CAN Newsletter 1*. Homerton College, Cambridge: CAN Continuation Project.
Caldwell, A. P. K. (1972) Motivation, emotional and interpersonal factors. In L. R. Chapman (ed.), *The Process of Learning Mathematics*. Oxford: Pergamon Press.
Cambridge Institute of Education (1985) *New Perspectives on the Mathematics Curriculum*. London: DES.
Carpenter, T. P. and Moser, J. M. (1982) The development of addition and subtraction problem-solving skills. In Carpenter, T. P., Moser, J. M. and Romberg, T. A., *Addition and Subtraction: A Cognitive Perspective*. Hillsdale: Erlbaum.
Carraher, T. N., Carraher, D. W. and Schliemann, A. D. (1985) Mathematics in the streets and in the schools. *British Journal of Developmental Psychology* **3**, 21–9.
Carter *et al.* (1981) *Mathematics in Biology*. Walton-on-Thames: Nelson.
Child, D. (1986) *Psychology and the Teacher* (4th edn). London: Holt, Rinehart & Winston.
Choat, E. (1978) *Children's Acquisition of Mathematics*. Windsor: NFER.
Claxton, G. (1984) *Live and Learn*. London: Harper & Row.
Cobb, P. (1987) Information-processing psychology and mathematics education—a constructivist perspective. *Journal of Mathematical Behaviour* **6**, 3–40.
Cockcroft, W. H. (1982) *Mathematics Counts*. London: HMSO.

Copeland, R. W. (1979) *How Children Learn Mathematics*. New York: Macmillan.
d'Ambrosio, U. (1985) Ethnomathematics and its place in the history and pedagogy of mathematics. *For the Learning of Mathematics* **5**(1), 44–8.
Dasen, P. R. (1972) Cross-cultural Piagetian research: a summary. *Journal of Cross-Cultural Psychology* **3**(1), 23–39.
Dasen, P. R. (ed.) (1977) *Piagetian Psychology: Cross-Cultural Contributions*. New York: Gardner Press.
Davis, R. B. (1966) Discovery in the teaching of mathematics. In L. S. Shulman and E. R. Keislar (eds), *Learning by Discovery: A Critical Appraisal*. Chicago: Rand McNally.
Davis, R. B. (1983) Complex mathematical cognition. In H. P. Ginsburg (ed.), *The Development of Mathematical Thinking*. New York: Academic Press.
Davis, R. B. (1984) *Learning Mathematics: The Cognitive Science Approach to Mathematics Education*. London: Croom Helm.
Dewey, J. (1910) *How We Think*. Boston, MA: Heath.
Dickson, L., Brown, M. and Gibson, O. (1984) *Children Learning Mathematics*. Eastbourne: Holt, Rinehart & Winston (Schools Council).
Dienes, Z. P. (1960) *Building Up Mathematics*. London: Hutchinson.
Dienes, Z. P. (1973) *The Six Stages in the Process of Learning Mathematics*. Windsor: NFER.
Donaldson, M. (1978) *Children's Minds*. Glasgow: Fontana/Collins.
Eliot, J. and Smith, I. M. (1983) *An International Directory of Spatial Tests*. Windsor: NFER–Nelson.
Esler, W. K. (1982) Physiological studies of the brain: implications for science teaching. *Journal of Research in Science Teaching* **19**, 795–803.
Farrell, M. A. and Farmer, W. A. (1980) *Systematic Instruction in Mathematics for the Middle and High School Years*. Reading, MA: Addison-Wesley.
Fennema, E. and Tartre, L. A. (1985) The use of spatial visualization in mathematics by girls and boys. *Journal for Research in Mathematics Education* **16**, 184–206.
Flanders, N. A. (1970) *Analyzing Teaching Behaviour*. Reading, MA: Addison-Wesley.
Fogelman, K. R. (1970) *Piagetian Tests for the Primary School*. Windsor: NFER.
Fox, L. H., Brody, L. and Tobin, D. (1980) *Women and the Mathematical Mystique*. Baltimore: Johns Hopkins University Press.
Furneaux, W. D. and Rees, R. (1978) The structure of mathematical ability. *British Journal of Psychology* **69**, 507–12.
Fuys, D., Geddes, D. and Tischler, R. (1988) *The van Hiele Model of Thinking in Geometry among Adolescents*. Reston, VA: National Council of Teachers of Mathematics.
Gagné, R. M. (1965, 1970, 1977, 1985) *The Conditions of Learning and Theory of Instruction*. New York: Holt, Rinehart & Winston.
Gagné, R. M. (1975) *Essentials of Learning for Instruction*. Hinsdale, IL: Dryden Press.
Gagné, R. M. and Briggs, L. J. (1974) *Principles of Instructional Design*. New York: Holt, Rinehart & Winston.
Gagné, R. M. and Brown, L. T. (1961) Some factors in the programming of conceptual learning. *Journal of Experimental Psychology* **62**, 313–21.
Gagné, R. M. and Smith, E. C. (1962) A study of the effects of verbalization on problem-solving. *Journal of Experimental Psychology* **63**, 12–18.
Gattegno, C. (1960) *Modern Mathematics with Numbers in Colour*. Reading: Educational Explorers.
Gay, J. and Cole, M. (1967) *The new mathematics and an old culture*. New York: Holt, Rinehart & Winston.
Gerdes, P. (1988) On culture, geometrical thinking and mathematics education. *Educational Studies in Mathematics* **19**, 137–62.
Getzels, J. W. and Jackson, P. W. (1962) *Creativity and Intelligence*. New York: Wiley.
Ginsburg, H. (1977) *Children's Arithmetic: The Learning Process*. New York: Van Nostrand.
Ginsburg, H. P. (ed.) (1983) *The Development of Mathematical Thinking*. New York: Academic Press.
Goldin, G. A. (1989) Constructivist epistemology and discovery learning in mathematics. In *Proceedings of the thirteenth annual conference of PME*.

Gross, T. F. (1985) *Cognitive Development.* Monterey, CA: Brooks/Cole.
Guilford, J. P. (1959) Three faces of intellect. *The American Psychologist* **14**, 469–79. (Also in Wiseman, S. (ed.) (1967) *Intelligence and Ability*. Harmondsworth: Penguin Books.)
Hadamard, J. (1945) *The Psychology of Invention in the Mathematical Field.* Princeton: Princeton University Press.
Hadow, W. H. (1931) *Report of the Consultative Committee on the Primary School.* London: HMSO.
Hardy, G. H. (1940) *A Mathematician's Apology*. Cambridge University Press.
Hart, K. (1980) A hierarchy of understanding in mathematics. In W. F. Archenhold *et al.* (eds), *Cognitive Development Research in Science and Mathematics.* University of Leeds Centre for Studies in Science Education.
Hart, K. M. (ed.) (1981) *Children's Understanding of Mathematics: 11–16*. London: John Murray.
Hart, K. M. (1984) *Ratio: Children's Strategies and Errors.* Windsor: NFER–Nelson.
Hartley, J. R. (1980) *Using the Computer to Study and Assist the Learning of Mathematics.* University of Leeds Computer-Based Learning Unit.
Harvey *et al.* (1982) *Language Teaching and Learning 6: Mathematics*. London: Ward Lock.
Heim, A. W. (1970) *AH4 Group Test of General Intelligence.* Windsor: NFER.
Hershkowitz, R. (1990) Psychological aspects of learning geometry. In P. Nesher and J. Kilpatrick, *Mathematics and Cognition.* Cambridge: Cambridge University Press.
Hill, C. C. (1979) *Problem-solving: Learning and Teaching.* London: Frances Pinter.
HMI (1979) *Mathematics 5–11*. London: HMSO.
HMI (1985) *Mathematics from 5 to 16.* London: HMSO.
Hodge, S. E. and Seed, M. L. (1972) *Statistics and Probability.* Glasgow, London, Edinburgh: Blackie, Chambers.
Holmes *et al.* (1980) *Teaching Statistics 11–16.* Slough: Foulsham (Schools Council).
Holt, J. (1969) *How Children Fail.* Harmondsworth: Penguin Books.
Hope, J. A. (1985) Unravelling the mysteries of expert mental calculation. *Educational Studies in Mathematics* **16**, 355–74.
Howard, R. W. (1987) *Concepts and Schemata.* London: Cassell.
Hudson, L. (1966) *Contrary Imaginations.* Harmondsworth: Penguin Books.
Hughes, E. R. (1980) Should we check children? In W. F. Archenhold *et al.* (eds), *Cognitive Development Research in Science and Mathematics.* University of Leeds Centre for Studies in Science Education.
Hughes, M. (1986) *Children and Number.* Oxford: Basil Blackwell.
Husen, T. (ed.) (1967) *International Study of Achievement in Mathematics*. London: Wiley.
Hutt, C. (1972) *Males and Females*. Harmondsworth: Penguin Books.
Jagger, J. M. (1985) A review of the research into the learning of mechanics. In A. Orton (ed.), *Studies in Mechanics Learning.* University of Leeds Centre for Studies in Science and Mathematics Education.
Joint Matriculation Board/Shell Centre for Mathematical Education (1984) *Problems with Patterns and Numbers: An O-level Module.* Manchester: Joint Matriculation Board.
Kamii, C. K. with DeClark, G. (1985) *Young Children Reinvent Arithmetic: Implications of Piaget's Theory*. New York: Teachers College Press.
Kamii, C. K. with Joseph, L. L. (1989) *Young Children Continue to Reinvent Arithmetic—2nd grade: Implications of Piaget's Theory*. New York: Teachers College Press.
Kane, R. B., Byrne, M. A. and Hater, M. A. (1974) *Helping Children Read Mathematics*. New York: American Book Co.
Karplus, R. and Peterson, R. W. (1970) Intellectual development beyond elementary school II: Ratio, a survey. *School Science and Mathematics* **70**, 813–20.
Katona, G. (1940) *Organizing and Memorizing.* New York: Columbia University Press.
Kelly, G. A. (1955) *The Psychology of Personal Constructs.* New York: Norton.
Kempa, R. F. and McGough, J. M. (1977) A study of attitudes towards mathematics in relation to selected student characteristics. *British Journal of Educational Psychology* **47**, 296–304.
Kilpatrick, J. (1985) Reflection and recursion. *Educational Studies in Mathematics* **16**, 1–26.

Krathwohl, D. R., Bloom, B. S. and Masia, B. B. (1964) *Taxonomy of Educational Objectives: The Affective Domain*. London: Longman.
Krutetskii, V. A. (1976) *The Psychology of Mathematical Abilities in Schoolchildren*. Chicago: University of Chicago Press.
Laborde, C. (1990) Language and mathematics. In P. Nesher and J. Kilpatrick, *Mathematics and Cognition*. Cambridge: Cambridge University Press.
Lancy, D. F. (1983) *Cross-Cultural Studies in Cognition and Mathematics*. New York: Academic Press.
Larcombe, A. (1985) *Mathematical Learning Difficulties in the Secondary School*. Milton Keynes: Open University Press.
Leder, G. (1985) Sex-related differences in mathematics: an overview. *Educational Studies in Mathematics* **16**, 304–9.
Lesh, R. and Landau, M. (1983) *Acquisition of Mathematics Concepts and Processes*. New York: Academic Press.
Lester, F. K. Jr. (1977) Ideas about problem-solving: a look at some psychological research. *The Arithmetic Teacher* **25**(2), 12–14.
Lindsay, P. H. and Norman, D. A. (1977) *Human Information Processing*. New York: Academic Press.
Lochhead, J. (1985) New horizons in educational development. In *Review of Research in Education*. Washington: American Educational Research Association.
Lovell, K. (1961) *The Growth of Basic Mathematical and Scientific Concepts in Children*. London: University of London Press.
Lovell, K. (1971a) Proportionality and probability. In M. F. Rosskopf, L. P. Steffé and S. Taback (eds), *Piagetian Cognitive-development Research and Mathematical Education*. Reston, VA: National Council of Teachers of Mathematics.
Lovell, K. (1971b) *The Growth of Understanding in Mathematics: Kindergarten Through Grade Three*. New York: Holt, Rinehart & Winston.
Lunzer, E. A. (1976) Towards an epistemological theory of mathematics learning. In E. A. Lunzer, A. W. Bell and C. M. Shiu, *Number and the World of Things*. University of Nottingham Shell Centre for Mathematical Education.
Lysaught, J. P. and Williams, C. M. (1963) *A Guide to Programmed Instruction*. New York: Wiley.
McIntosh, A. (1977) When will they ever learn? *Forum* **19**, 92–5.
MacNamara, E. A. (1990) Subitizing and addition of number: a study of young children learning mathematics. M.Ed. thesis, University of Leeds.
Mager, R. F. (1975) *Preparing Instructional Objectives* (issued 1962 as *Preparing Objectives for Programmed Instruction*). Belmont, CA: Fearon.
Manchester Mathematics Group (1970) *A Structural Approach to Mathematics: Unit 2*. London: Rupert Hart-Davis.
Mathematical Association (1923) *The Teaching of Geometry in Schools*. London: Bell.
Mathematical Association (1970) *Primary Mathematics—A Further Report*. Mathematical Association.
Matthews, G. (1964) *Matrices 1*. London: Edward Arnold.
Midlands Mathematical Experiment (1964) *Report 1962–63*. London: Harrap.
Midlands Mathematical Experiment (1970) *GCE Volume II Part B*. London: Harrap.
Miller, G. A. (1956) The magical number seven, plus or minus two: some limits on our capacity for processing information. *Psychological Review* **63**, 81–97.
Mitchelmore, M. C. (May 1980) Three-dimensional geometrical drawing in three cultures. *Educational Studies in Mathematics* **11**, 205–16.
Morris, R. W. (1974) Linguistic problems encountered by contemporary curriculum development projects in mathematics. In UNESCO, *Interactions Between Linguistics and Mathematical Education*. UNESCO/CEDO/ICMI.
Newell, A. and Simon, H. A. (1972) *Human Problem Solving*. Englewood Cliffs, NJ: Prentice-Hall.
Novak, J. D. (1977) *A Theory of Education*. Ithaca: Cornell University Press.
Novak, J. D. (1980) Methodological issues in investigating meaningful learning. In W. F.

Archenhold *et al.* (eds), *Cognitive Development Research in Science and Mathematics*. University of Leeds Centre for Studies in Science Education.
Novak, J. D. and Gowin, D. Bob (1984) *Learning How to Learn*. Cambridge: Cambridge University Press.
Nuffield Mathematics Project (1967a) *I Do, and I Understand*. Edinburgh, London, New York: Chambers, John Murray, Wiley.
Nuffield Mathematics Project (1967b) *Pictorial Representation*. Edinburgh, London, New York: Chambers, John Murray, Wiley.
Nuffield Mathematics Project (1969) *Computation and Structure 4*. Edinburgh, London: Chambers, John Murray.
Nuffield Mathematics Project (1970, 1973) *Checking Up I, II, III*. Edinburgh, London, New York: Chambers, John Murray, Wiley.
Nuffield Maths 5–11 (1983) *Nuffield Maths 6 Teacher's Handbook*. Harlow: Longman.
Orton, A. (1970) A cross-sectional study of the development of the mathematical concept of a function in secondary schoolchildren of average and above average ability. M.Ed. thesis, University of Leeds.
Orton, A. (1971) Teaching about functions in the secondary school. *Mathematics Teaching* **57**, 45–9.
Orton, A. (1977) A mathematics concepts workshop. *Mathematical Education for Teaching* **2**(4), 19–27.
Orton, A. (1980) A cross-sectional study of the understanding of elementary calculus in adolescents and young adults. Ph.D. thesis, University of Leeds.
Orton, A. (1983a) Students' understanding of integration. *Educational Studies in Mathematics* **14**, 1–18.
Orton, A. (1983b) Students' understanding of differentiation. *Educational Studies in Mathematics* **14**, 235–50.
Orton, A. (1985) *Studies in Mechanics Learning*. University of Leeds Centre for Studies in Science and Mathematics Education.
Papert, S. (1980) *Mindstorms*. Brighton: Harvester Press.
Peak Mathematics (see Brighouse, A., Godber, D. and Patilla, P.).
Piaget, J. (1973) *The Child's Conception of the World*. London: Paladin.
Pimm, D. (1987) *Speaking Mathematically*. London: Routledge.
Plowden, B. (1967) *Children and their Primary Schools*. London: HMSO.
Poincaré, H. (1924) Mathematical creation. Reprinted in P. E. Vernon (ed.) (1970) *Creativity*. Harmondsworth: Penguin Books.
Polya, G. (1945) *How to Solve It*. Princeton: Princeton University Press.
Polya, G. (1954) *Mathematics and Plausible Reasoning*. Princeton: Princeton University Press.
Polya, G. (1962) *Mathematical Discovery*. New York: Wiley.
Rees, R. M. (1973) *Mathematics in Further Education*. London: Hutchinson.
Rees, R. (1974) An investigation of some common mathematical difficulties experienced by students. *Mathematics in School* **3**(1), 25–7.
Rees, R. (1981) Mathematically gifted pupils: some findings from exploratory studies of mathematical abilities. *Mathematics in School* **10**(3), 20–3.
Renwick, E. M. (1935) *The Case Against Arithmetic*. London: Simpkin Marshall.
Resnick, L. B. and Ford, W. W. (1984) *The Psychology of Mathematics for Instruction*. Hillsdale: Lawrence Erlbaum.
Romberg, T. A. (1982) An emerging paradigm for research on addition and subtraction skills. In T. P. Carpenter, J. M. Moser and T. A. Romberg. *Addition and Subtraction: A Cognitive Perspective*. Hillsdale: Erlbaum.
Roper, T. (1985) Students' understanding of selected mechanics concepts. In A. Orton (ed.), *Studies in Mechanics Learning*. University of Leeds Centre for Studies in Science and Mathematics Education.
Rosskopf, M. F., Steffé, L. P. and Taback, S. (eds) (1971) *Piagetian Cognitive-development Research and Mathematical Education*. Reston, VA: National Council of Teachers of Mathematics.

Russell, S. (1983) *Factors Influencing the Choice of Advanced Level Mathematics by Boys and Girls*. University of Leeds Centre for Studies in Science and Mathematics Education.
Saxe, G. B. and Posner, J. (1983) The development of numerical cognition: cross-cultural perspectives. In H. P. Ginsburg, *The Development of Mathematical Thinking*. New York: Academic Press.
Scandura, J. M. and Wells, J. M. (1967) Advance organizers in learning abstract mathematics. *American Educational Research Journal* **4**, 295–301.
Scheerer, M. (1963) Problem-solving. *Scientific American* **208**(4), 118–28.
Schliemann, A. D. (1984) Mathematics among carpentry apprentices: implications for school teaching. In *Mathematics for All*. UNESCO Document Series No. 20.
Schools Council (1965) *Mathematics in Primary Schools*. London: HMSO.
Seaborne, P. L. (1975) *An Introduction to the Dienes Mathematics Programme*. London: University 9f London Press.
Shuard, H. B. (1982a) Differences in mathematical performance between girls and boys. In W. H. Cockcroft, *Mathematics Counts*. London: HMSO.
Shuard, H. B. (1982b) Reading and learning in mathematics. In R. Harvey *et al.*, *Language Teaching and Learning 6: Mathematics*. London: Ward Lock.
Shuard, H. and Rothery, A. (eds) (1984) *Children Reading Mathematics*. London: John Murray.
Shulman, L. S. (1970) Psychology and mathematics education. In E. G. Begle (ed.), *Mathematics Education*. Chicago: NSSE.
Skemp, R. R. (1964) *Understanding Mathematics: Teacher's Notes for Book 1*. London: University of London Press.
Skemp, R. R. (1971) *The Psychology of Learning Mathematics*. Harmondsworth: Penguin Books.
Skemp, R. R. (1976) Relational understanding and instrumental understanding. *Mathematics Teaching* **77**, 20–6.
Skemp, R. R. (1982) Communicating mathematics: surface structures and deep structures. *Visible Language* **XVI**, 281–8.
Skinner, B. F. (1954) The science of learning and the art of teaching. *Harvard Educational Review* **24**, 86–97.
Skinner, B. F. (1961) Teaching machines. *Scientific American* **205**(5), 90–102.
Smith, I. M. (1964) *Spatial Ability: Its Educational and Social Significance*. London: University of London Press.
Solomon, Y. (1989) *The Practice of Mathematics*. London: Routledge.
Springer, S. P. and Deutsch, G. (1981) *Left Brain, Right Brain*. San Francisco: Freeman.
Starkey, P. and Gelman, R. (1982) The development of addition and subtraction abilities prior to formal schooling in arithmetic. In T. P. Carpenter, J. M. Moser and T. A. Romberg, *Addition and Subtraction: A Cognitive Perspective*. Hillsdale: Erlbaum.
Stern, C. with Stern, M. B. (1953) *Children Discover Arithmetic*. London: Harrap.
Stewart, J. (1985) Cognitive science and science education. *European Journal of Science Education* **7**, 1–17.
Stewart, J. and Atkin, J. (1982) Information processing psychology. *Journal of Research in Science Teaching* **19**, 321–32.
Sutton, C. (ed.) (1981) *Communicating in the Classroom*. London: Hodder & Stoughton.
Suydam, M. N. and Weaver, J. F. (1977) Research on problem-solving: implications for elementary school classrooms. *The Arithmetic Teacher* **25**(2), 40–2.
Tait, K., Hartley, J. R. and Anderson, R. C. (1973) Feedback procedures in computer-assisted arithmetic instruction. *British Journal of Educational Psychology* **43**, 161–71.
Thorndike, E. L. (1922) *The Psychology of Arithmetic*. New York: Macmillan.
Torbe, M. and Shuard, H. (1982) Mathematics and language. In R. Harvey *et al.*, *Language Teaching and Learning 6: Mathematics*. London: Ward Lock.
UNESCO (1974) *Interactions Between Linguistics and Mathematical Education*. UNESCO/CEDO/ICMI.
Vergnaud, G. (1982) A classification of cognitive tasks and operations of thought involved in addition and subtraction problems. In T. P. Carpenter, J. M. Moser and T. A. Romberg. *Addition and Subtraction: A Cognitive Perspective*. Hillsdale: Erlbaum.

Vergnaud, G. (1990) Epistemology and psychology of mathematics education. In P. Nesher and J. Kilpatrick, *Mathematics and Cognition*. Cambridge: Cambridge University Press.
Vernon, P. E. (1950) *The Structure of Human Abilities*. London: Methuen.
von Glasersfeld, E. (1987) Learning as a constructive activity. In C. Janvier (ed.), *Problems of Representation in the Teaching and Learning of Mathematics*. Hillsdale: Erlbaum.
Vygotsky, L. S. (1962) *Thought and Language*. New York: MIT Press/Wiley.
Walkup, L. E. (1965) Creativity in science through visualization. *Perceptual and Motor Skills* **21**(1), 35–41.
Wall, W. D. (1965) Learning to think. In W. R. Niblett (ed.), *How and Why Do We Learn?* London: Faber & Faber.
Wertheimer, M. (1961) *Productive Thinking*. London: Tavistock Publications.
Wickelgren, W. A. (1974) *How to Solve Problems*. San Francisco: Freeman.
Williams, J. S. (1985) Using equipment in teaching mechanics. In A. Orton (ed.), *Studies in Mechanics Learning*. University of Leeds Centre for Studies in Science and Mathematics Education.
Wilson, B. (1981) *Cultural Contexts in Science and Mathematics Education*. University of Leeds Centre for Studies in Science and Mathematics Education.
Witkin *et al.* (1977) Field-dependent and field-independent cognitive styles and their educational implications. *Review of Educational Research* **47**(1), 1–64.
Wood, R. (1977) Cable's comparison factor: is this where girls' troubles start? *Mathematics in Schools* **6**(4), 18–21.
Wrigley, J. (1963) Some programmes for research. In F. W. Land (ed.), *New Approaches to Mathematics Teaching*. London: Macmillan.
Young, R. G. (1966) *Sets: A Programmed Course*. London: Methuen.

Name Index

Subject Index